Thomas Reid and the Problem of Secondary Qualities

Edinburgh Studies in Scottish Philosophy
Series Editor: Gordon Graham,
Center for the Study of Scottish Philosophy, Princeton Theological Seminary

Scottish philosophy through the ages

This new series will cover the full range of Scottish philosophy over five centuries – from the medieval period through the Reformation and Enlightenment periods, to the nineteenth and early twentieth centuries.

The series will publish innovative studies on major figures and themes. It also aims to stimulate new work in less intensively studied areas, by a new generation of philosophers and intellectual historians. The books will combine historical sensitivity and philosophical substance which will serve to cast new light on the rich intellectual inheritance of Scottish philosophy.

Books available
Adam Smith and Rousseau: Ethics, Politics, Economics, edited by Maria Pia Paganelli, Dennis C. Rasmussen and Craig Smith
Thomas Reid and the Problem of Secondary Qualities, Christopher A. Shrock
Hume's Sceptical Enlightenment, Ryu Susato

Books forthcoming
Imagination in Hume's Philosophy: The Canvas of the Mind, Timothy M. Costelloe
Adam Ferguson and the Idea of Civil Society: Moral Science in the Scottish Enlightenment, Craig Smith
Essays on Hume, Smith and the Scottish Enlightenment, Christopher Berry
Eighteenth-Century Scottish Aesthetics: Not Just a Matter of Taste, Rachel Zuckert

www.edinburghuniversitypress.com/series/essp

Thomas Reid and the Problem of Secondary Qualities

Christopher A. Shrock

EDINBURGH
University Press

Edinburgh University Press is one of the leading university presses in the UK. We publish academic books and journals in our selected subject areas across the humanities and social sciences, combining cutting-edge scholarship with high editorial and production values to produce academic works of lasting importance. For more information visit our website: edinburghuniversitypress.com

© Christopher A. Shrock, 2017

Edinburgh University Press Ltd
The Tun – Holyrood Road, 12(2f) Jackson's Entry, Edinburgh EH8 8PJ

Typeset in 11/13 Adobe Sabon by
IDSUK (DataConnection) Ltd, and
printed and bound in Great Britain by
CPI Group (UK) Ltd, Croydon CR0 4YY

A CIP record for this book is available from the British Library

ISBN 978 1 4744 1784 6 (hardback)
ISBN 978 1 4744 1785 3 (webready PDF)
ISBN 978 1 4744 1786 0 (epub)

The right of Christopher A. Shrock to be identified as the author of this work has been asserted in accordance with the Copyright, Designs and Patents Act 1988, and the Copyright and Related Rights Regulations 2003 (SI No. 2498).

Contents

Acknowledgements

My interest in Thomas Reid and his theory of primary and secondary qualities began in 2008, with a graduate seminar taught by Todd Buras at Baylor University. That course led to a series of presentations and short publications on the nature of Reid's primary/secondary distinction, his analyses of various secondary quality species, and eventually Reid's relevance to contemporary perception theory. While I was working out my interpretation of Reid, I received feedback and other helps from Baylor professors and classmates, plus significant financial support for conference travel from the university. Todd directed my dissertation on Reid and proved a brilliant mentor, even after I left Waco in 2010 for a position at the Oklahoma School of Science and Mathematics. At OSSM, I benefitted from the encouragement and patience of key administrators Drs Edna Manning, Ken Lease and Frank Wang. Dr Manning also helped me obtain checkout privileges at the University of Oklahoma library. In 2012 Todd suggested that I develop my research into a book, and Oklahoma Christian University's Beam Library provided the academic and other resources to realise that goal. My wife, Danielle, has been wonderful since the start.

Series Editor's Preface

It is widely acknowledged that the Scottish Enlightenment of the eighteenth century was one of the most fertile periods in British intellectual history, and that philosophy was the jewel in its crown. Yet, vibrant though this period was, it occurred within a long history that began with the creation of the Scottish universities in the fifteenth century. It also stretched into the nineteenth and twentieth centuries for as long as those universities continued to be a culturally distinctive and socially connected system of education and inquiry.

While the Scottish Enlightenment remains fertile ground for philosophical and historical investigation, these other four centuries of philosophy also warrant intellectual exploration. The purpose of this series is to maintain outstanding scholarly study of great thinkers like David Hume, Adam Smith and Thomas Reid, alongside sustained exploration of the less familiar figures who preceded them, and the impressive company of Scottish philosophers, once celebrated, now neglected, who followed them.

Gordon Graham

PART I

Why Secondary Qualities are a Problem

Consider your perceptions as you read this book. What do you see? At least a book. But be more specific. What about the book – which of its properties? No doubt, you see and feel its physical dimensions. Perhaps you also perceive the contrast of black lettering on white pages, or the luminescent contrast of a tablet. When I read, I can often make out features of my hands along the edges of the pages and a little of the surrounding environment. What do you feel? The smoothness of the paper or its lukewarm temperature? I tend to leave my books in the car and have to fetch them again. In the winter, I love the way the pages feel after sitting in the cold for a while, icy to the touch. What do you smell – book glue, your coffee? What do you hear? What things are around you, and what properties do they exhibit to your senses? Take notice of the details.

Now suppose you learn that at least some of these properties exist only in your mind? For now, I'm not talking about the size, shape or texture of the book. Of course, those things are really there. But what if today's best physicists and evolutionary psychologists discovered that glue has no smell, that letters and pages are not black and white, and that lukewarm-ness is merely a psychological state or chemical reaction in your brain? Strictly speaking, every time you have ever thought of colours, smells, sounds or heat as existing in a real world apart from a human perceiver's mind, you have been mistaken. As far as external reality is concerned, there is no such thing as the colour of a paint sample, no sound of your niece's piano recital, no taste of my grandmother's angel food cake.

Perhaps there are those who would suffer no scandal if this were the case. Pop science and relativistic sophistry have trained them to expect illusions of this sort. But if you're like me, then such a discovery would be nothing short of shocking. I'm not offended at the bare possibility that I might fall victim to an illusion. But if colours, sounds and tastes do not exist, then my senses have deceived me my entire life. The world suddenly seems dark and mysterious, and I am angered by my own inability to comprehend it.

If you don't immediately sympathise with my reaction to this hypothetical case, consider the following line of questioning. If you learned that colours, smells and sounds were mere illusions, then what would you make of your perceptual abilities as a whole? Would you have reason to rely on them at all? Could you trust yourself, for example, to judge music and visual art, or wine and the culinary arts? If you once considered J. S. Bach or Georgia

O'Keeffe or Wolfgang Puck to possess talent or extraordinary skill, you might be forced to admit that their gifts have significance only in relation to subjective sensations or emotions. They only matter in your mind. And if someone else can't stand Bach's Prelude and Fugue in C major or O'Keefe's *Summer Days* or Puck's salmon pizza, then you're wrong to think their tastes dysfunctional. Even if you don't find this disturbing, are there not times when you take yourself to be observing things objectively? Science relies heavily on perception. Would failures of perception undermine your faith in the very science that gave you knowledge of those failures? I confess, it would mine. And where will we turn for knowledge once science, or common sense itself, gives way? If you can sympathise with such concerns, then you have begun to understand my interest in what I call the Problem of Secondary Qualities.

The seeming discrepancy between appearance and reality meets philosophers, both past and present, at every turn. The problem is so complex, so widely dispersed through philosophical and scientific literature, that no one could address all of it in a single work. I confine myself to a tiny corner of the debate. That is the conflict between common sense perceptual judgments and claims among some philosophers and scientists concerning the natures of so-called secondary qualities – colour, sound, smell, taste, and heat. Many argue that these properties are purely mental, non-existent without perceivers to perceive them. For these individuals, secondary qualities are relations between perceivers and physical objects, mere mental experiences, or sensory misrepresentations. This last description construes secondary qualities as epistemically sinister, as if common sense would have us believe an entire system of falsehoods. It is enough to threaten a person's confidence in common sense altogether. This is the Problem of Secondary Qualities.

Among philosophers, debate over which beliefs count as commonsensical is often heated. But pretty much everyone agrees that, according to common sense, immediate objects of sense perceptions – including secondary qualities – are mind-independent, objective and physical. Philosophers recognise this view as Direct Realism. Direct Realism is a thesis about objects of perception. More formally:

Direct Realism: Physical entities and certain physical qualities are among the immediate objects of human sense perceptions.

According to Direct Realism, you perceive mind-external objects, and your perceptions of those objects are not mediated by other perceptions of mental, subjective or spiritual entities. Furthermore, according to Direct Realism, objects of sense perceptions are not sensations, qualia, sense-data or perceiver-dependent relations.

If the notions of mediate and immediate objects are unfamiliar to you, consider a game in which causal chains play a feature role, like billiards. Say you're playing a game of Eight-Ball, and you want to sink the four, which is sitting conveniently near a corner pocket. But the path from the cue ball to the four is blocked by the six. You're frustrated that you cannot connect the cue ball and the four directly, but it's no reason for despair. You call out, 'Combo!' and line up on the six. If you have a bit of skill or luck, the cue will knock the six into the four and drop the latter into the pocket. In this case, the six is your immediate target, and the four is a mediate target. You target the four by or in virtue of targeting the six.

Mediated communication is also a familiar phenomenon, although not by that name. The clearest examples involve one person using another to communicate. Consider a deaf person and a hearing person conversing by means of a translator. The deaf person signs immediately to the translator who changes the medium of the message to vocal speech and passes it on to the hearing person, and vice versa. The deaf person communicates with the hearing person, but that communication is mediated by the translator. The translator is the immediate recipient of the message, and the hearing person the mediate recipient. One may easily envision other examples by considering circumstances where immediate communication is difficult. In the 1988 comedy *Crocodile Dundee II*, the two main characters discover their mutual commitment to their relationship on a long and crowded subway platform, where their conversation is mediated by a series of helpful commuters.

Direct Realism says that humans perceive physical objects immediately, as supposed by common sense. Opposed to Direct Realism are two rival theories, Indirect Realism and Idealism. For them, the Problem of Secondary Qualities is cause to believe that cognitive contact with the world outside the human mind is vastly different than you may have imagined. Indirect Realism, for example, says that physical objects and their qualities are not direct or immediate objects of perception, but perceptions of them are mediated by certain mental objects.

> Indirect Realism: Physical entities and physical qualities are among the objects of sense perceptions, but the immediate objects of human perceptions are mental entities and mental qualities.

According to Indirect Realism, perceptions of physical things are had in virtue of perceiving mental or subjective entities. You might directly see a rectangular sense-datum, for example, and in virtue of this perception you might also see a photograph, albeit indirectly. In this case, you do not see the photograph immediately. Rather, your perception of the photograph is mediated by your perception of the sense-datum, much as the earlier mentioned shooting of the four ball was mediated by the shooting of the six and the conversation between the deaf and hearing individuals was mediated by the translator. You take credit for seeing the photograph in virtue of seeing the sense-datum, just as you take credit for sinking the four in virtue of having hit the six with the cue ball and the hearing person takes credit for communicating with the deaf person in virtue of having relayed her message through the translator.

In-virtue-of perceptions are open to both Direct and Indirect Realists. Both might say that you see a certain desk in virtue of seeing the top of the desk; and you see the top of the desk in virtue of seeing the surface of its top; and you see the surface of its top in virtue of seeing the parts of that surface that are not covered by computers, books and the usual office clutter. This chain eventually finds its terminus at the immediate object(s) of your perception. The point of contention between Direct and Indirect Realists concerns the nature of that terminal object. What is it? Direct Realism answers that it is an external, mind-independent, physical thing, like a desk or a photograph. Indirect Realism says that it is a subjective, mental entity – a sense-datum, qualia, sensum or idea. Following Thomas Reid and the tradition preceding him, I prefer to speak of these hypothetical mental objects of perception as ideas. But please don't confuse this technical use of the term with the everyday use in which ideas are closer to concepts, notions or thoughts.

Philosophers' talk of perceiving properties parallels their talk of perceiving objects. They speak of properties as objects of perception. Consider a penny viewed at an angle. Direct Realism claims that you see the penny's roundness directly, despite the fact that your sense experience is different than when looking at the penny from its zenith. Indirect Realism agrees that you see the penny's roundness, but it disagrees that you see it directly. Instead, it claims, you perceive the penny's roundness in virtue of perceiving an elliptical idea. After all, when viewing the penny from an angle, you see something elliptical, but the penny is circular. So the elliptical object must be a mediating idea that represents the physical penny.

Among the competing theories of perception, Idealism is the naysayer. It says nay to physical objects of perception and nay to in-virtue-of perceptions.

> Idealism: Mental entities and certain mental properties are the only objects of human sense perceptions.

According to Idealism, Indirect Realism is correct that immediate objects of perception are ideas. But Idealism denies that ideas can cognitively connect perceivers to physical things. According to this view, the elliptical, mental, penny-like idea is all you see and, for all you know, there is no physical penny.

Much of this book focuses on a metaphysical distinction between entities that exist in their own right and entities that depend on perceiving subjects. Apparently, there is no perfect terminology to be applied here. 'Perceiver-independent' and 'perceiver-dependent' are good but cumbersome to readers. 'Real' and 'unreal' appear in early modern writings but have different meanings for contemporary philosophers. 'Internal' and 'external' seem vague and metaphorical. And 'objective' and 'subjective' result in peculiar locutions about 'objective objects' as opposed to 'subjective objects'. So it is with some trepidation that I adopt 'mental' and 'physical', knowing that these terms may ill-dispose a few readers. The best I can do is to plead for charity. I ask the reader to treat my terms 'physical' and 'mental' as marking the same distinction that could have been expressed in different, although equally objectionable language. In particular, I do not mean to be dismissive of physicalism about the mind. Physicalism about the mind is not a target in this work, and perhaps Reid's approach to secondary qualities could be translated into physicalist terms. My aim is to defend Direct Realism against Indirect Realism and Idealism, a goal that most physicalists will appreciate.

I mention Indirect Realism and Idealism in conjunction with Direct Realism because the Problem of Secondary Qualities is often more a matter of choosing a favourite than accepting or rejecting a lone theory. This book means to answer the Problem of Secondary Qualities, but it would be impossible to stop there. The assessment of Direct Realism has implications for other theories and philosophical issues. So before giving the Problem of Secondary Qualities its formal introduction, consider a few reasons why you might favour Direct Realism as well.

I

Why Direct Realism?

If you're new to the philosophy of perception, you may wonder why Direct Realism is worth defending against Indirect Realism or Idealism. Tough question! A comprehensive answer would encompass this book and many others. For starters, though, consider some advantages that Direct Realism boasts over its rivals. First, it is the recognised common sense position. Odds are that you already believe it, and it fits best with your other beliefs – like that you've felt a friend's handshake or seen the ocean without feeling or seeing any mental objects in between. If Direct Realism turns out false, then you may have to revise many, many other beliefs. Perhaps you'd be inclined to think that reality is out to get you! Second, Indirect Realism has trouble accounting for the in-virtue-of relation that connects perceivers to the physical world. How can seeing an idea count as seeing the physical world? Finally, Indirect Realism and Idealism usually show up as promised solutions to the infamous Problems of Illusion and Hallucination, but how viable are they as solutions? The Problems of Illusion and Hallucination cast doubt on human abilities to acquire perceptual knowledge. Indirect Realism and Idealism are supposed to secure that knowledge by making perceptions always veridical, at least in a qualified sense. But introducing evanescent ideas as objects of perception, in addition to or instead of common sense physical objects, seems a poor approach to ensuring veridical perception. It only makes things worse. This chapter spells out one advantage of Direct Realism over Idealism, one over Indirect Realism, and a self-undermining feature common to Direct Realism's rivals in order to justify this book's default acceptance of Direct Realism.

Direct Realism as Common Sense

Human thinkers acquire beliefs in various ways: perception, reason, memory, testimony and perhaps others. Philosophers call these ways epistemic faculties. If all is well, you find your faculties in harmony. You remember leaving your keys on the counter, but to be certain you ask your friend whether they are in fact on the counter. She testifies that they are, and when you go to retrieve them, you perceive them there for yourself. All your faculties converge on the truth that your keys are on the counter. That's a typical case.

Idealism posits a great struggle between the faculties of reason and perception. Sense perception inclines beliefs toward a world of physical objects that dwell beyond perceivers' own minds. Reason, however, reduces objects of perception to ideas only, at least according to Idealism. How does it resolve the discrepancy? By giving reason authority over the inclinations of perception. Maybe perception just isn't that informative. It's up to cool reason to find the truth. But perception may not be quieted so easily. If Idealism speaks for cool reason, common sense offers a more fiery voice on behalf of perception. Samuel Johnson famously objected to George Berkeley's Idealism by kicking a stone along a road to regular shouts of 'I refute it thus!'[1] Silly as his argument seems, Johnson's thesis is clear: there is a physical world, and it's right here. For him, Idealism elicits a visceral reaction; it creates such dissonance in his mind.

Berkeley hopes that his Idealism, on which everything is either a perceiving spirit or immaterial object of perception, will be received as commonsensical.[2] His hope, it seems, is vain. Idealism failed to impress Johnson, but consider also Berkeley's writings themselves. In one place he denies that Idealism conflicts with common sense, but elsewhere he acknowledges a natural tendency to view the objects of perceptions as mind-independent. There Berkeley links the tendency to regard perceptual objects as mind-independent with the involuntary nature of perceptions and perceptual objects. You don't get to choose whether or not you perceive this object, so it doesn't seem to be part of you.[3] Another prominent Idealist, David Hume, remarks eloquently,

> Men are carried, by a natural instinct or prepossession, to repose faith in their senses; and that, without any reasoning, or even almost before the use of reason, we always suppose an external

universe, which depends not on our perception but would exist, though we and every sensible creature were absent or annihilated.[4]

Berkeley and Hume regard faith in sense perception as mistaken and easily undermined, but also as common sense and natural. So their brand of Idealism introduces a dangerous tension among epistemic faculties. To those who wish to pursue truth with seriousness, their brand of Idealism advocates the submission of perception, and perhaps other faculties, to reason and philosophical speculation.

On the other hand, Direct Realism tries to align reason with the natural inclinations of sense perception. On a daily basis, you probably trust reason and perception to yield true beliefs and to cooperate with memory and other capacities. In my life, this cooperation has yielded considerable successes, enabling my learning and my safety. Unless you've suffered some surprising and wildly unfortunate circumstances, your story is probably the same, receiving valuable information from perception, memory, testimony and reason alike. An odd sensation or difficult environmental circumstances may occasionally lead to false perceptual beliefs. But errors also arise in logic and arithmetic, which are reason's purview. Sometimes reasoning corrects perceptual errors, but perception also checks reason. Science requires both raw data and mathematical coherence. So what obvious reason is there to privilege reason over sense perception? Nothing comes to mind. As Thomas Reid says in response to the external world sceptic,

> Why, Sir, should I believe the faculty of reason more than that of perception; they came both out of the same shop, and were made by the same artist; and if he puts one piece of false ware into my hands, what should hinder him from putting another?[5]

Reid's contemporary, David Hume divides objects of thought into relations of ideas and matters of fact. Then he demands that his reader justify the latter in terms of the former.[6] Why pit reason against its fellow faculties? Direct Realism preserves a common sense cognitive harmony, while Idealism introduces an unpalatable tension among the faculties.

G. E. Moore's famed attack on Idealism illustrates the unyielding power of sense perceptions. He privileges certain perceptual beliefs in an effort to turn the Idealist framework on its head.

Moore begins with the claims, 'Here is a hand. And here is another'. Then, he says that, because hands are physical objects, at least two physical objects exist, which means that there is a physical world beyond his mind.[7] Moore's basic gist challenges the foundations of Idealism by pointing out the implausibility of its conclusion.

You might accuse Moore of begging the question. Whether or not there are hands and other physical objects is the thing in dispute, not a self-evident truth! If Idealism is right, then there are not physical hands but hand-ish amalgamations of ideas, sense-data or qualia. Doesn't Moore miss the point? Maybe not. Direct Realist Michael Huemer offers a more subtle understanding of Moore's argument.[8] According to him, Moore changes the nature of the dispute between Idealists and Direct Realists. Of course, Idealism won't accept Moore's line, 'Here is a [physical] hand', but the argument cuts both ways. The perception-based conviction, 'Here is a [physical] hand', undermines Idealism as well. You can see physical hands or adhere to Idealism, but not both. For Moore, belief in his own hands effectively overwhelms his confidence in any particular theory of perception. Naturally and rationally, he rejects those theories which deny him his most certain beliefs. As a revisionary metaphysics, Idealism owes common sense thinkers sustained successful arguments before demanding that they relinquish natural and deeply committed perceptual beliefs.

Noah Lemnos finds the common thread in Moore and Reid. As he explains, Moore is on to something about the power of perceptual beliefs.[9] Thomas Reid characterises some of them as irresistible and expresses helplessness when it comes to determining his beliefs. About perceptual beliefs, Reid says,

> If [perceptual beliefs are judgements of nature], as I apprehend [they are], it will be impossible to shake off those opinions, and we must yield to them at last, though we struggle hard to get rid of them. If we could, by a determined obstinacy, shake off the principles of our nature, this is not to act the philosopher, but the fool or the madman.[10]
>
> Shall we then throw off this [perceptual] belief [that hardness is a quality of external bodies], as having no foundation in reason? Alas! it is not in our power; it triumphs over reason, and laughs at all the arguments of a philosopher. Even the author of the *Treatise of human nature*, though he saw no reason for this belief, but many against it, could hardly conquer it in his speculative and solitary

moments; at other times he fairly yielded to it, and confesses that he found himself under a necessity to do so.[11]

Nature enslaves Reid's mind to the deliverances of perception, and Reid accuses Hume, Idealist author of the *Treatise of Human Nature*, of capitulating to them as well. Hume himself openly acknowledges perception's tremendous force towards beliefs in physical things,

> When we believe any thing of external existence, or suppose an object to exist a moment after it is no longer perceived, this belief is nothing but a sentiment of the same kind. Our author insists upon several other sceptical topics; and upon the whole concludes, that we assent to our faculties, and employ our reason only because we cannot help it. Philosophy would render us entirely Pyrrhonian, were not nature too strong for it.[12]

In eighteenth-century Scotland, for both Direct Realists and Idealists, then, perceptual beliefs were often unshakable. In the twentieth century, the same sentiment appears in P. F. Strawson: 'The grip that commonsense non-reductive realism has on our ordinary thinking . . . so thoroughly permeates our consciousness that even those who are intellectually convinced of its falsity remain subject to its power.'[13] If Moore's critic calls his belief in physical hands into question, Reid, Hume and Strawson seem to place it beyond question. Moore couldn't change it if he wanted to.

How can Idealism ask one to give up one's irresistible beliefs? It's unfair to hold one responsible for what is beyond one's control. As the Kantian puts it, 'Ought implies can'. If you can't change your perceptual beliefs in physical things, then are you not, in some important sense, justified in holding them? Reid, who sees voluntariness and responsibility as intimately connected, offers the following example, involving an unintentionally disobedient servant, to motivate that connection:

> Suppose a servant, through negligence and inattention, mistakes the orders given him by his master, and, from this mistake, does what he was ordered not to do. It is commonly said that culpable ignorance does not excuse a fault: This decision is inaccurate, because it does not shew where the fault lies: The fault was solely in that inattention or negligence, which was the occasion of his mistake: There was no subsequent fault.

This becomes evident, when we vary the case so far as to suppose, that he was unavoidably led into the mistake without any fault on his part. His mistake is now invincible, and, in the opinion of all moralists, takes away all blame; yet this new case supposes no change, but in the cause of his mistake. His subsequent conduct was the same in both cases. The fault therefore lay solely in the negligence and inattention which was the cause of his mistake.

The axiom, That invincible ignorance takes away all blame, is only a particular case of the general axiom, That there can be no moral obligation to what is impossible; the former is grounded upon the latter, and can have no other foundation.[14]

For Reid, involuntary beliefs are always justified. If your natural belief that you perceive certain physical objects is irresistible – that is, obtained in the course of upholding your day-to-day ethical and epistemic responsibilities – then you have no obligation to change it, even if Idealism makes an impressive case for itself.

Surely, conflicting with justified beliefs is a strike against a theory. If perceptual beliefs in physical things are invincible, then Idealism can never fully reconcile with them. For the sake of consistency, then, perhaps it's worth taking Reid's advice to 'make a virtue of necessity' by avoiding Idealism.[15] The question is whether Direct Realism can withstand the philosophical scrutiny.

Indirect Realism and the Veil of Ignorance

Recall that Direct and Indirect Realists subscribe to in-virtue-of perceivings, in which one sees a book in virtue of seeing its cover, and one sees its cover in virtue of seeing its spine, and so on. They differ on the nature of the terminal or immediate object of perception. Direct Realism says that the immediate object is physical. Indirect Realism says that the immediate object is mental, like an idea or sense-datum.

One question for Indirect Realism, then, is how to cross the gap from mental to physical. How is it that perceiving an idea causes or counts as perceiving a physical thing? Seeing a book by seeing its cover suggests that perceptions can track part–whole relations – seeing the book by seeing part of the book. Perhaps there are other relations on which in-virtue-of perceivings piggyback. A gloved handshake seems like a case where one perceives the hand of another despite the intervening layers of clothing. But are ideas or sense-data related to physical entities such that perceiving the former brings about perceiving the latter? This is a more challenging case. I have not been able to find a reason why perceptions of mere ideas should cause or count as perceptions of physical bodies. The ideas are not parts or properties of physical things, nor do they cover them as clothing. It is not clear whether or how an idea might transmit information about physical things to perceivers, nor why the perception of one should count as evidence for the existence of the other. How does Indirect Realism bridge the chasm?

Challenging the epistemic step from the perceiving of ideas to the perceiving of physical things is an old criticism. Berkeley points to this problem, likely targeting Indirect Realist John Locke:

> As for our senses, by them we have the knowledge only of our sensations, ideas, or those things that are immediately perceived by sense, call them what you will: but they do not inform us that things exist without the mind, or unperceived, like to those which are perceived. . . . But what reason can induce us to believe the existence of bodies without the mind, from what we perceive, since the very patrons of matter themselves do not pretend, there is any necessary connexion betwixt them and our ideas?
>
> . . . perhaps it may be thought easier to conceive and explain the manner of their production, by supposing external bodies in their likeness rather than otherwise; and so it might be at least probable there are such things as bodies that excite their ideas in our minds. But neither can this be said; for though we give the materialists their external bodies, they by their own confession are never the nearer knowing how our ideas are produced . . . Hence it is evident the production of ideas or sensations in our minds, can be no reason why we should suppose matter or corporeal substances, since it is acknowledged to remain equally inexplicable with, or without this supposition.[16]

Reid finds Berkeley's criticism of Indirect Realism convincing. He does not see how ideas could count as evidence for physical things. Berkeley, Reid says, 'hath proved, beyond the possibility of reply, that we cannot by reasoning infer the existence of matter from our sensations'.[17] The same sentiment appears nearly three hundred years later in the writings of contemporary Idealist John Foster:

> It seems just obvious that, if the sensible object immediately before his mind is a sense-quale, and if the only other relevant aspect of

his current psychological condition is the accompanying experimental interpretation, then the subject's awareness stops at the quale itself and does not make contact with anything external.[18]

Indirect Realism needs an account of the in-virtue-of relationship or process that brings you from the immediate perceiving of an idea to the perceiving of a physical thing. Otherwise, ideas form a veil, not a window, between your mind and the physical world.

For the sake of fairness, consider two proposals for establishing the in-virtue-of connection between idea and physical object of perception. First, Locke attempts to base the connection on something called resemblance. On the resemblance theory, perceptions of a physical thing succeed when you perceive an idea with properties that resemble the properties of the physical thing. It works best with a special group of properties that Locke calls primary qualities – size, figure, motion and a few others.[19] You immediately perceive some ideas, and this counts as a mediated perception of a certain physical entity, because the immediately perceived ideas resemble the physical properties of size, figure, motion and so forth. That is, you perceive a physical thing or its properties in virtue of perceiving ideas, because the ideas resemble the physical thing's properties.

Reid, who quite reasonably takes Locke's ideas to be sensations, says that this is absurd. He writes,

When I . . . compare [primary qualities and sensations] together, it appears to me clear as day-light, that the former are not of kin to the latter, nor resemble them in any one feature. They are as unlike, yea as certainly and manifestly unlike, as pain is to the point of a sword.[20]

Take an idea, like pain. Compare it to the physical properties of a correlative physical object, like a sword. Any resemblance? None whatsoever! The resemblance theory, says Reid, cannot account for knowledge of the physical world, since ideas do not in fact resemble physical things in any significant respect and so cannot serve as evidence for them. Pain is not like the point of a sword.

From across the English Channel, Jean-Jacques Rousseau submits another proposal for establishing the in-virtue-of relation between ideas and physical things, suggesting that ideas might take physical things as objects in a more general sense.[21] The perceived idea is an object of perception, and the physical object is an object of the perceived idea. But the object-taking does not depend on resemblance. Thoughts and writings exhibit intentionality like this. An essay on Thomas Reid takes Reid as an object, since it is of Reid. And the activity of writing the essay takes the essay as an object, since it is the writing of the essay. But the activity of writing does not resemble an essay, nor does the essay resemble Reid himself. Likewise, perhaps immediate perceptions are of ideas, and ideas are of physical things, and this chain of intentionality means that perceptions can be of physical things, although only mediately. The obvious downside of such a view is the lack of explanation. How do intentionality relations arise if not through resemblance?

Again supposing ideas to be sensations, Reid finds reasons to reject Rousseau's more general sort of Indirect Realism. For him, sensations take no objects, other than perhaps themselves.[22] Lack of object-taking, Reid thinks, distinguishes sensations from other operations of the mind, like memories or reasonings. Reid's point has some intuitive appeal. After all, pains are pains and pleasures are pleasures if one experiences them, no matter what causal or intentionality relations might obtain between them and physical things or other mental operations. The notion of sensation loses nothing by failing to have relations of intentionality. Furthermore, Rousseau's proposal doesn't say exactly how to bridge the epistemic gap. A pain's being of a sword does not show how the pain grants perceptual knowledge of the sword's physical properties. So even Rousseau's generalised Indirect Realism has problems.

Ultimately, Reid finds Indirect Realism an impossible position. There just isn't a good way to make the cognitive move from ideas to physical objects of perception. Reid almost laments, 'Our sensations have no resemblance to external objects; nor can we discover, by our reason, any necessary connection between the existence of the former, and that of the latter'.[23] And in another place,

> And as the feeling [caused by a touching a hard object] hath no similitude to hardness, so neither can our reason perceive the least tie or connection between them; nor will the logician ever be able to show a reason why we should conclude hardness from this feeling, rather than softness, or any other quality whatsoever. But in reality all mankind are led by their constitution to conclude hardness from this feeling.[24]

Reid interpreter John Greco uses this passage to build an argument for external world scepticism on Reid's behalf. He dubs it the 'No Good Inference' argument. It says that, if you are supposed to learn about the external world by reasoning from your ideas or sensations, then you're is in trouble. There is no good inference from sensations to physical objects. And, if this is the case, then none of us can have knowledge of physical things, at least not by perception.[25]

Without an adequate explanation of the connection between immediate perceptions of ideas and mediated perceptions of physical things, Indirect Realism remains highly unsatisfying. Introducing hypothetical ideas as the immediate objects of perception leads more naturally to Idealism. But as illustrated in the previous section, Idealism has its problems too.

Can Rival Theories Solve the Problem of Illusion?

The final advantage Direct Realism has over Idealism and Indirect Realism concerns the motives for positing mental entities as immediate objects of sense perception. Both Indirect Realism and Idealism use immediately perceived mental entities in order to solve certain theoretical problems with illusions and hallucinations. In particular, they aim to explain perceptual errors, or perception-based inclinations toward false beliefs, in terms of veridical immediate perceptions. In these cases, mental entities, like ideas or sense-data, present themselves fully transparent perceptual objects, perspicuous and accessible to their perceiver. You understand them in their entirety just by perceiving them.

For example, if you take a half-submerged stick in water to look bent, then the illusion is partially explained by the immediate perception of a bent idea. After all, something is bent, and it isn't the stick! Or if you hallucinate so that you seem see a brown rectangular table, this approach says that even if there isn't a table matching your experience, you do in fact see a brown rectangular idea. The immediate perception is accurate. By adopting ideas as immediate objects of perception, Indirect Realists and Idealists make perceivers immune to illusions and hallucinations, at least at the immediate level. Howard Robinson, who features prominently in the next chapter, defends sense-data for always being as they seem.[26] Other philosophers who connect ideas with perceptual infallibility are legion, including H. H. Price, M. G. F. Martin and A. J. Ayer.[27] Perhaps their stance is inherited from Locke, who writes, 'Whoever reflects on what passes in his own Mind cannot miss it', and, 'For let any Idea be as it will, it can be no other but such as the Mind perceives it to be'.[28]

If ideas are supposed to guarantee immediate perceptions, then Indirect Realism and Idealism face two threatening objections. The first simply challenges the claim that ideas are always as they seem. Keith Lehrer considers a case in which an itch seems to be a pain or vice versa.[29] If there can be a genuine confusion about the nature of the sensation and one can misperceive or fail to perceive an idea, then ideas just aren't doing the work for which they were hired. If ideas make immediate perceptions infallible, then there should be no confusion or vulnerability to error.

While Lehrer's pain–itch case is problematic, Indirect Realism and Idealism face a second, more extreme problem as well. It is that immediate objects of perception do not seem like mental entities. Suppose that a naive perceiver directs her eyes towards a certain white desk. The Indirect Realist says that she sees, immediately, a rectangular, white idea. But what does she take herself to see? If asked, she responds, 'A rectangular, white table'. Suppose the Indirect Realist explains to her that only the immediate object of her perception matters, and 'immediate' means that perceptual object in virtue of which she sees whatever things she sees. Perhaps he tries an example: you see a certain car in virtue of seeing its wheels and body, and you see its body in virtue of seeing a door and fender, and you see the door and fender in virtue of seeing the paint on the door and fender, and so on. The final link in this chain, he says, is the immediate object of perception. Now, how does she respond, having got clarification on the vocabulary? Does she say that she sees, immediately, a white, rectangular idea? Or does she say that the immediate object of her perception is such-and-such part of the table? The answer seems obvious. The naive perceiver still says that she sees some part or quality of the table, rather than a hypothetical mental entity.

There is a problem here. It seems to the subject that she immediately perceives a physical object or part of a physical object. But if her immediate object of perception is an idea, then either the idea seems not to be an object of perception or the idea seems to be a physical entity. The first option seems intuitive. If the naive perceiver perceives ideas, then she somehow skips them when analysing her perceptions. She just isn't reflective enough to find the true immediate object of perception. But Idealist David Hume suggests the other route, that the perceived ideas seem physical to the perceiver. He writes, 'It seems also evident, that, when men follow this blind and powerful instinct of nature, they always suppose the very images to be the external [physical] objects.'[30] For Hume, ideas seems like physical objects or properties! Either way, ideas aren't as they seem. Either they are immediate objects of perception and seem not to be or they are mental and seem physical. Regardless, ideas are not wholly accessible to perceivers.

Since Indirect Realists and Idealists posit ideas in order to account for hallucinations and illusions in terms of veridical perceptions, this is a huge problem for them. Mistaking ideas for physical things and failing to recognise ideas as objects of perception are not minor confusions but major misclassifications. Ideas, if they exist, are not at all as they seem. Therefore, they generate additional illusions rather than explaining illusions and hallucinations. This makes them theoretically self-defeating and at least as illusory as physical objects of perception. So Indirect Realism and Idealism lose their motive for suggesting ideas as immediate objects of perception.

Direct Realism faces challenges, but it also possesses some advantages over its rivals. Unlike Idealism, it is the common sense view and accords with humanity's natural, perhaps irresistible, beliefs in physical objects. Moreover, Direct Realism preserves the natural harmony among epistemic faculties. Unlike Indirect Realism, it puts perceivers in cognitive contact with physical objects of perception rather than walling them off behind ideas. Finally, Direct Realism does not ramify philosophical problems relating to illusion and hallucination by positing mental objects of perception that do not deliver on their promises to ensure the accuracy of immediate perceptions. Overall, there are some excellent reasons to invest some hope in Direct Realism.

Notes

1. Boswell, *Life of Johnson*, p. 130.
2. Berkeley, *A Treatise Concerning the Principles of Human Knowledge*, 1.1/89.
3. Berkeley, *A Treatise Concerning the Principles of Human Knowledge*, 1.56/122.
4. Hume, *An Enquiry Concerning Human Understanding*, 12.1/104.

5. Reid, *Inquiry into the Human Mind*, 6.20/169.
6. Hume, *An Enquiry Concerning Human Understanding*, 4/108–18.
7. Moore, 'Proof of an External World'.
8. Huemer, *Skepticism and the Veil of Perception*, pp. 31–45. See also Lemos, *Common Sense*, pp. 85–104.
9. Lemos, *Common Sense*, pp. 13–23.
10. Reid, *Inquiry into the Human Mind*, 2.7/37.
11. Reid, *Inquiry into the Human Mind*, 5.2/58.
12. Hume, *Treatise on Human Nature*, Abstract/414.
13. Strawson, 'Perception and its Objects', p. 53.
14. Reid, *Essays on the Active Powers of Man*, 4.7/238.
15. Reid, *Inquiry into the Human Mind*, 5.7/68.
16. Berkeley, *A Treatise Concerning the Principles of Human Knowledge*, 1.18–19/109.
17. Reid, *Inquiry into the Human Mind*, 5.7/70, 5.4/61.
18. Foster, *The Nature of Perception*, p. 203.
19. Locke, *An Essay Concerning Human Understanding*, 2.8.15/137.
20. Reid, *Inquiry into the Human Mind*, 5.7/68.
21. Rousseau, *Emile*, VI/211–356.
22. Reid, *Essays on the Intellectual Powers of Man*, 1.1/36; Buras, 'The Nature of Sensations in Reid', pp. 221–38.
23. Reid, *Inquiry into the Human Mind*, 6.21/176. Repeated in Reid, *Essays on the Intellectual Powers of Man*, 3.7/289–90.
24. Reid, *Inquiry into the Human Mind*, 5.5/64.
25. Greco, John, 'Reid's Reply to the Skeptic', p. 143.
26. Robinson, *Perception*, p. 32.
27. Huemer, 'Sense-Data'; Price, *Perception*, p. 3; Martin, 'Beyond Dispute', pp. 218–19; Ayer, 'Has Austin Refuted the Sense-Datum Theory?', p. 129; Ayer, *The Foundations of Empirical Knowledge*, pp. 3–11.
28. Locke, *An Essay Concerning Human Understanding*, 2.9.1/143, 2.29.5/364.
29. Lehrer, *Knowledge*, pp. 95–9.
30. Hume, *Enquiry Concerning Human Understanding*, 12.1/104.

General Exposition of the Problem of Secondary Qualities

The most influential attacks on Direct Realism fall into a family of objections known collectively as the Argument from Illusion or Problem of Illusion. These begin by considering a case in which, the Argument from Illusion alleges, the true object of perception differs from the physical thing that you naively take yourself to be perceiving. When you direct your eyes towards the straight stick in water, what you see is obviously bent. But the physical stick is not bent. Therefore, what you see, at least immediately, is not the physical stick. Perhaps it is a mental thing, like an idea or sense-datum. From here, the Argument typically employs what I call a spreading principle to deny Direct Realism altogether. The stick case is, in crucial respects, qualitatively similar to other perceptions of physical objects, so surely all other immediate objects of perception are mental too. It concludes that Direct Realism is wrong: human perceivers do not immediately perceive physical things or properties.

The Problem of Secondary Qualities is an Argument from Illusion. It says that secondary qualities (colour, smell, sound, taste and heat) are possessed by objects of perception but not by physical things. You can perceive them, but they only exist in your mind, as mental properties or objects. Then the Problem considers other properties, like length, solidity, shape, visual shape and texture, in comparison with the secondary qualities. It points out a qualitative continuity between the phenomenological and experiential aspects of secondary quality perceptions and perceptions of other qualities. Consider what it's like to run your hand over a glass table or velvet upholstery, to squeeze a marble or lump of wet clay. Think about the visual appearance of an unusually shaped drinking glass. These experiences do not seem radically different from those of colour, smell, sound, taste or heat. Secondary quality perceptions share many features, including epistemic features, with perceptions of other properties. Why think that you can see a book's shape immediately when you know that its colour, which seems to inhabit the limits of its shape, belongs to an idea? Why think that you can feel the surface of your coffee cup directly when you know that its warmth, which seems to be permeating that surface, dwells only in your mind? Better to think that secondary quality perceptions mirror perceptions of other objects and properties.

As a rule, early modern natural philosophers understand secondary qualities as illusory. Their arguments, however, seem to focus on only one secondary quality – heat, colour or taste – rather than taking them as a group. Therefore, I turn to a general argument from Howard Robinson's more recent work *Perception*:

> Science has shown that physical objects do not possess secondary qualities intrinsically. As they are clearly possessed by that of which we are aware in perception, that of which we are aware in perception is not the physical object itself. The only plausible way to understand the relation between physical objects and secondary qualities is to think of the objects as possessing dispositions to produce the qualities in us as properties of our sense-data.[1]

In this remarkably short presentation, Robinson contrasts Indirect Realism's satisfaction of a key desideratum with Direct Realism's supposed failure. Direct Realism, he says, cannot account for the fact that subjects perceive secondary qualities. Secondary qualities are mental, and, according to Direct Realism, objects of perception are physical. So, Robinson argues, Indirect Realism turns out to have the upper hand in these cases. What a marvellous claim!

It's critical to the aims of this book to have a more formal analysis of Robinson's argument. To begin, there is a cryptic assumption at work here that is elsewhere explicit in Robinson's text. It is the Spreading Principle, the statement that generalises from the illusory case to all instances of perception. Robinson's broadest version of such a principle goes, 'There is such a continuity between those cases in which objects appear other than they actually are and cases of veridical perception that the same analysis of perception must apply to both'.[2] In its defence, Robinson asserts,

> There is no absolute distinction between a state of tiredness in
> which things look slightly less clear and a less tired state, or
> between accurate vision and very slight short-sightedness; nor
> probably, is there such a thing as absolutely accurate perception
> of colour, rather a slight variation between persons. It is, there-
> fore, very implausible to say that some of these cases involve direct
> apprehension of an external object and in the others of a sense-
> datum. So the argument generalises easily.[3]

Consider the phenomenological and physiological differences
between your own illusions and veridical perceptions, especially
the milder cases like the partially submerged stick in water or a
round tower that looks rectangular from a distance. Wouldn't it
be strange if the natures of perceptions of dry sticks were wildly
different from the natures of perceptions of partially submerged
sticks? How plausible is it that from ten meters you perceive the
round tower immediately, but from ten kilometres you perceive
a rectangular idea instead? Such inconsistencies strike Robinson,
and perhaps you too, as suspiciously ad hoc. When you pull the
stick out of the water, for example, you probably don't notice a
bent idea exiting your virtual field to be replaced by a physical
stick. So why suppose that secondary quality perceptions differ in
kind from perceptions of other properties? Perceptions of colours,
sounds and tastes form a phenomenological whole along with figure, size, motion and many others.

While Robinson offers a compelling case for his general
spreading principle, a narrower version may suit the Problem
of Secondary Qualities better. Consider a restricted version of
Robinson's spreading principle that highlights the connection
between illusoriness in secondary quality perceptions and omits
the bit about sense-data:

> Secondary Quality Spreading Principle (SP): If there are immediate
> objects of perception that possess secondary qualities and those
> objects are not physical, then no immediate objects of perception
> are physical.

This restriction does no violence to the original principle because it
commits Robinson to less, just as a mathematical corollary can be
simultaneously less informative and more interesting than its root
axiom. Robinson himself makes the same type of move elsewhere,
like in his justification of a hallucination spreading principle.[4]

Besides the Spreading Principle, consider two additional claims to which Robinson is explicitly committed. The first is just the fact of secondary quality perceptions:

> Secondary Quality Observation Claim (OC): Some immediate objects of perception possess secondary qualities.

The second Robinson takes as a deliverance of science:

> Secondary Quality Non-Physicality Thesis (NPT): If an object is physical, then it does not possess secondary qualities.

From these three premises – SP, OC and NPT – one can deduce the falsity of Direct Realism. Hence the Problem of Secondary Qualities.

To navigate from premises to conclusion, consider an immediate object of perception that possesses secondary qualities, one of the objects mentioned in OC, the Observation Claim. The object can't be physical, according to NPT, the Non-Physicality Thesis. So there must be at least one immediate object of perception that possesses secondary qualities but isn't physical. And SP, the Spreading Principle, says that the existence of this one non-physical immediate object of perception shows that no immediate object of perception is physical. Which yields Robinson's desired conclusion, that Direct Realism is false. Robinson goes on to argue for the existence of sense-data, but there is no reason to bother with that here. The damage is done. Despite objections against Idealism and Indirect Realism in the previous chapter, the argument is valid. If SP, OC and NPT are true, then Direct Realism is false.

Experienced philosophers know well the principle exercise of the next chapter, probing a valid argument for weak premises. The problem with the Problem of Secondary Qualities is that a Direct Realist can't accept SP, OC and NPT and remain a Direct Realist without succumbing to accusations of irrationality. Philosophical good form demands a rejection of at least one premise if a valid argument's conclusion becomes intolerable. If Direct Realism is correct, then something must be wrong with SP, OC or NPT. But which one? There are excellent reasons for believing all three. Even as a critic of the Problem of Secondary Qualities, I accept the Spreading Principle and the Observation Claim. I defend them in the next chapter. Ultimately, this means denying the third premise,

NPT, that physical objects do not possess secondary qualities. There turn out to be good reasons to regard secondary qualities as physical properties, even contrary to the opinions of many physicists, neuroscientists and fellow philosophers.

Part II of this book proposes a theory of secondary qualities on which they may be identified with physical, scientific properties. Scientists may avoid traditional secondary quality names, but they do make implicit references to secondary qualities. If heat is kinetic energy, minute molecular vibrations and the occasional runaway particle, then chemistry textbooks can omit discussions of 'heat' and instead discuss 'kinetic energy', veiling their treatment of the secondary quality with scientific vocabulary. Similarly, you could tell a story about your brother without ever mentioning the fact that he is your brother, but it would be about your brother nonetheless. Secondary qualities are in fact scientific and physical. If I'm right about this, then the Problem of Secondary Qualities disappears since NPT is false.

You may read this chapter and conclude that this discussion is about the language appropriate to secondary qualities. Don't be misled. It's not merely an issue of semantics. The Problem of Secondary Qualities concerns the nature of human sense perceptions. If the Problem is correct, then common sense trust in perception is misguided, and direct cognitive contact with the physical world is impossible. However, with the viability of an identity theory of secondary qualities, there is hope. Common sense has not lied to us perceivers. Rather philosophers have misunderstood certain nuances of perception. Perhaps the gap between appearance and reality is a mere crack or fissure rather than a gulf or chasm.

Notes

1. Robinson, *Perception*, p. 59.
2. Robinson, *Perception*, pp. 56–8.
3. Robinson, *Perception*, p. 57.
4. Robinson, *Perception*, pp. 87–8.

3

Why Direct Realism Needs Objective Secondary Qualities

The first chapter suggested some reasons to favour Direct Realism over Indirect Realism and Idealism. Direct Realism has some common sense appeal. The other theories face troubling challenges, including the illusoriness of ideas. Chapter 2, however, showed that Direct Realism cannot rationally stand alongside the three premises of the Problem of Secondary Qualities: the Secondary Quality Spreading Principle (SP), the Secondary Quality Observation Claim (OC) and the Secondary Quality Non-Physicality Thesis (NPT). Something must give. This chapter considers SP, OC and NPT more carefully in hopes of pinpointing some weakness that can save Direct Realism.

Surprisingly, the most vulnerable premise in the Problem of Secondary Qualities turns out to be NPT, the supposed deliverance from science that physical objects do not possess secondary qualities. Some philosophers may cringe at the thought of challenging scientific dogma, but the present chapter explains the rationale behind this move. SP and OC hold up under scrutiny, and NPT isn't quite as scientific as you might expect.

The Secondary Quality Spreading Principle (SP)

The first premise, the Secondary Quality Spreading Principle, says that if there are immediate objects of perception that possess secondary qualities and these are not physical, then no immediate objects of perception are physical. One could easily mistake it for the argument's weak point. SP seems finicky, preoccupied with secondary qualities in particular, and unmotivated. So what if perceiving secondary qualities means perceiving ideas or sense-data? Aren't secondary qualities exceptional? Why extrapolate an analysis of secondary quality perceptions to perception generally?

SP holds its ground surprisingly well. Recall that SP entered the discussion as a special case of Howard Robinson's more general spreading principle. Robinson based his claim on causal and physiological continuities between veridical perceptions and slight illusions. Your visual perception of a clock's face, for example, seems pretty much the same whether you observe it in a bleary-eyed stupor or with the wakeful clarity of a late-morning cup of coffee. Similarly, George Berkeley bases a version of SP on the phenomenological continuity between secondary and other properties. Witness the following exchange between Berkeley's characters, Hylas and Philonous:

> Phil. Again, is it your opinion that colours are at a distance?
> Hyl. It must be acknowledged, they are only in the mind.
> Phil. But do not colours appear to the eye as coexisting in the same place with extension and figures?
> Hyl. They do.
> Phil. How can you then conclude from sight, that figures exist without, when you acknowledge colours do not: the sensible appearance being the very same with regard to both?
> Hyl. I know not what to answer.[1]

Secondary qualities co-appear with other physical properties. You experience them together, says Berkeley, so they stand or fall together. In fact, Berkeley goes as far as to say that one cannot even conceptualise a physical body as existing without its secondary qualities:

> But I desire any one to reflect and try, whether he can by any abstraction of thought, conceive the extension and motion of a body, without all other sensible qualities. For my own part, I see evidently that it is not in my power to frame an idea of a body extended and moved, but I must withal give it some colour or other sensible quality which is acknowledged to exist only in the mind. In short, extension, figure, and motion, abstracted from all other qualities, are inconceivable. Where therefore the other sensible qualities are, there must these be also, to wit, in the mind and no where else.[2]

Berkeley means that he cannot abstract secondary qualities away from the primaries in the Lockean sense of considering a hypothetical object that lacks secondary qualities while possessing primary qualities. That is, he cannot form a conception of an object that has primary qualities but lacks secondary qualities, as Galileo seems to do when he considers a world without secondary qualities.[3] David Hume seconds Berkeley's opinion, saying, 'After the exclusion of colours, sounds, heat and cold from the rank of external existences, there remains nothing, which can afford us a just and consistent idea of body'.[4] Even in imagination, secondary qualities conjoin with primary qualities and other seemingly objective properties. This is important because, for Berkeley and Hume, conceivability is the mark of possibility.[5] It is inconceivable that a body be extended without also bearing some colour or temperature. He cannot separate primary and secondary qualities via abstraction, so an account of one spills into the other.

John Foster treats secondary qualities as merely the tip of the iceberg. Quickly spreading his version of Sense-Datum Theory from secondary qualities to all perceivable objects, he writes,

> This is just the point, already mentioned, about the status of the secondary qualities – that science seems to show that such qualities as colour, sound, flavour, and odour are nothing more, in the physical items themselves, than powers to affect human sense-experience, together with the primary structures on which these powers are grounded. But, importantly, it also extends to the primary structures themselves. For, even in respect of spatial patterning, how things sensibly appear to the ordinary visual and tactual observer is not, except in broad outline, the same as how things turn out in the light of microscopic and sub-microscopic investigation. The conclusion to be drawn seems to be that our perception of the physical world is non-veridical on a global scale, and so almost entirely beyond the reach of presentationalist [the view that physical objects are present in or partially constitutive of our perceptual experiences] treatment.[6]

The Problem, if allowed to fester, ramifies into full-blown scepticism concerning the external physical world in the twenty-first century as well as in the eighteenth.

This is especially true for secondary qualities, Foster thinks, since secondary qualities occupy so much of human perceptual life. If these are illusory, then 'illusion will be ubiquitous'.[7] There is also a phenomenological continuity, or at least a coupling, between secondary and other qualities. Colour appears with extension, heat with hardness. Co-experienced qualities, it seems, should receive similar philosophical treatments. So, as it turns out, SP has more appeal than one might suppose.

Consider two alternatives to SP. First, Behaviourism suggests a theory of perception on which there are no ideas but secondary qualities nevertheless depend on perceivers for their existence in a peculiar way that makes SP less plausible. Second, Disjunctivism attempts a more straightforward approach, maintaining OC and NPT while rejecting SP. Unfortunately, neither Behaviourism nor Disjunctivism preserves the common sense attachment to Direct Realism sought since Chapter 1. So this section concludes with an endorsement of SP and recommends seeking elsewhere for weaknesses in the Problem of Secondary Qualities.

Behaviourism

Behaviourist George Pitcher is among the few philosophers who explicitly deny the Secondary Quality Spreading Principle. Behaviourism says that the field of psychology, and talk about mental states and operations, should be accounted for in terms of the physical behaviours of individual organisms. The theory has fallen out of favour in the twenty-first century, but Pitcher is worth mentioning because of his tremendous influence on philosophers of perception since the 1970s as well as his specific attention to secondary qualities. Pitcher rejects SP by making an exception for secondary qualities. Secondary qualities, he suggests, depend on perceivers for their existence, while other perceivable properties remain physical and independent. In his own words,

> The way people's perceptual apparatus happens to be constructed determines to a much greater degree what colors things are said to be than it does what shapes things are said to be. Colors are in fact largely dependent on the nature of our visual equipment.[8]

You should be somewhat surprised at Pitcher's characterisation of colours, since Pitcher is a self-proclaimed Direct Realist and Direct Realists typically don't make such sharp distinctions between secondary qualities and other perceivable properties.

For Pitcher, perceptions are properly caused dispositions to behave in certain ways. Consider the example of seeing a tree. You see the tree because sunlight reflects off the tree, affects your eyes in a certain way, and puts your brain into a certain state. That brain state disposes you to do things like climb the tree, stand in its shade on a hot day, or walk around it rather than into it. According to Pitcher, acquiring such a dispositional state just is perceiving the tree.

Since Pitcher's perceptions are just dispositions to behave in certain ways, what you perceive and whether you perceive it veridically depends on how well you're disposed to accomplish certain tasks. For instance, if you see a circle and wish to trace it, you might use a compass. Setting one leg in the circle's centre and swinging the other around, you find that it easily conforms to the perimeter. You perceive the circle veridically because you're disposed to use appropriate actions and tools to trace the shape successfully. On the other hand, if you seem to see a circle but in reality it is a square, then you can't trace it with a compass. You see the square, but your compass-using behaviour fails to reflect the square's non-circularity. Your purposes get thwarted by your ill-suited behavioural dispositions. So, for Pitcher, this perception is non-veridical, an illusion. And it gets evaluated as an illusion because of the 'punishing experiences' you suffer, the lack of fit between your purposes and your actions.[9]

Regarding colour perceptions, Pitcher limits the relevant behavioural dispositions to only two types: matching and discriminatory.[10] For things to look the same colour, he claims, means that you become prone to matching those things on the basis of apparent colour. Likewise, two things look differently coloured if and only if you can sort them into different colour categories. Perhaps other behavioural dispositions attach to colour, but they aren't defining in the way that matching and discriminating are. For example, most men are especially attracted to women in red.[11]

Given this simple account of what it is to see colours, Pitcher goes on to note that colour perceptions, dispositions to match and discriminate, depend on one's anatomy in ways that other perceptions do not. If your visual organs were adjusted, even slightly, you might have notably different colour matching and discriminatory behaviours. At the same time, your perceptions would be veridical on Pitcher's account, because your altered sorting dispositions would not lead to significant punishing experiences. If you were sorting fruits by colour, you might disagree with someone about putting the red grapes with the radishes and tomatoes or with the beetroot and eggplants. But you won't get the appalling results of trying to trace a square with a compass. Pitcher, therefore, insists on counting abnormal colour perceptions as veridical.

If perceivers adopt different sorting dispositions in response to the same object and the different dispositions nevertheless count as veridical, then, Pitcher suggests, the natures of the perceptual objects must depend somehow on their perceivers. So he sets colours and perhaps all secondary qualities apart as subjective or relational, unlike other perceivable properties, which are objective because of their potential to bring about punishing experiences when misperceived. Sequestering secondary qualities as especially subjective constitutes Pitcher's grounds for denying SP. Contrary to the Problem of Secondary Qualities and SP, secondary qualities are special cases and so do not threaten Direct Realism generally. Direct Realism is vindicated.

Or perhaps not. Clever as Pitcher's objection to SP is, it presents two critical problems. First, he can't maintain the boundary between secondary qualities and other perceivable properties. Pitcher regards colour as subjective because slight anatomical alterations can yield radically different colour matching and discriminatory behaviours without generating severe punishing experiences. But compare colour to an obviously objective property, like weight or size.

To see how slight anatomical and physiological differences can result in very different behaviours, consider a simple case. Suppose that your blood sugar were low and you felt weak, then you would perceive objects as heavier on Pitcher's account. This is so because you would adopt different behavioural dispositions. You might use a wheelbarrow or cart to move something that you would normally carry. You incur no punishing experiences, so, according to Pitcher's earlier line, your perceptions are veridical. Perhaps weight, like colour, depends on perceivers. But surely not. You can measure weight with a scale. It's an objective property determined by gravitational forces.

Size presents a similar problem. For example, George Berkeley imagines a mite regarding its own foot as a body of 'considerable dimension'.[12] The mite's behaviours with regard to its foot, like using it to crawl, differ from those of a larger creature, like an ape – not only because the foot belongs to the mite rather than the ape! Neither mite nor the ape incurs punishing experiences for its behaviours. The mite treats the foot as mites do, and the ape ignores the mite's foot until the whole mite becomes a snack. So both perceive the foot veridically. On Pitcher's analysis of colour perceptions, differences in non-punishing perception-based behavioural dispositions reveal a perceivable property's subjectivity. Thus, on the Behaviourist approach, the size of the foot may depend on the perceiver as much as its colour. But, like weight, surely size is an objective property, even if Berkeley concludes that it isn't.

The weight and size examples undermine Pitcher's move from perceptual-physiological sensitivity to subjectivity of a perceived property. Colour isn't subjective just because behavioural dispositions towards it change with physiological attributes. Colour isn't exceptional in this regard. And if colour isn't exceptional, Pitcher loses his grounds for stopping SP.

Consider also whether Behaviourism is truly a species of Direct Realism. Direct Realism regards physical properties and physical things as immediate perceptual objects. Behaviourism excludes mental properties and mental things as immediate perceptual objects. But ridding the world of ideas and sense data doesn't automatically account for perceivers' abilities to see, feel and smell physical objects. How, then, does Behaviourism identify immediate perceptual objects? Are its perceptual objects also Direct Realism's objects? Not obviously. In fact, it's difficult to see how Behaviourism can pick out a perceptual object. For example, try tracing a behavioural disposition to its causal origins. Return again to the perception of the tree. Your behavioural disposition to stand in the tree's shade arises because of a causal chain that begins with the shining sun and ends in your mind or brain. Light from the sun reflects off the tree, which stimulates your eyes, which sets off a reaction in your nervous system, which changes your brain state, and so on. By hypothesis, the tree is the object of perception and part of the causal chain. But can Behaviourism distinguish it as the immediate perceptual object? Sunlight initiates the causal process. Reflected light exerts immediate causal influence on your body. Within your body, immediate causes of your behavioural dispositions occur in your brain. You're equally responsive to sun, sunlight, tree, reflected light, eyes, nerves and brain. So what makes the tree the object of perception as opposed to one of these other objects or events? Behaviourism offers no obvious answer.

When the intellectual battle concerns the nature of perceptual objects, it's hard to see what ground Behaviourism can gain for

Direct Realism. Pitcher's promise to separate secondary qualities from other perceivable properties falls through, because Behaviourism has trouble identifying perceptual objects, which is a basic aim of the present discussion.

A Disjunctivist Objection to the Spreading Principle

Behaviourism couldn't beat the Spreading Principle. But mightn't there be other approaches for excepting secondary quality perceptions from Direct Realism's general account of perception? One school of thought with resources for distinguishing veridical perceptions from illusions and hallucinations is Disjunctivism.

Disjunctivism has little to say about secondary qualities in particular, but it does suggest a direct attack on the Spreading Principle. It says that veridical perceptions, illusions and hallucinations share no common nature. Despite their phenomenal, physiological and perhaps causal similarities, arguments from analogies between illusions and veridical perceptions don't hold.[13] Direct Realist versions of Disjunctivism say that objects of perception are physical but consider the natures of illusions and hallucinations open for debate. Authentic perceptions just have different natures. So, even if secondary quality perceptions turn out to be illusory or take mental entities as objects, this fact carries no implications for perception or perceptual objects in general.

Certain Disjunctivists may challenge OC or NPT, but, for the sake of argument, consider a bare-bones Disjunctivism that accepts these as well as Robinson's contention that the possessors of secondary qualities must be ideas. Moreover, suppose that this version of Disjunctivism maintains the common sense view that some (perhaps most) perceivable properties belong to physical things, thus rendering the Problem of Secondary Qualities a local disturbance on a vast conceptual landscape.

How might Robinson or another defender of SP respond to this bare-bones Disjunctivism, which concurs with OC and NPT but not SP? Recall Robinson's earlier emphasis on the qualitative continuity between veridical and illusory scenarios. He points it out along environmental and physiological dimensions. The physiological state of a very tired, bleary-eyed person experiencing an illusion differs only slightly from a less tired person with veridical perceptions. How short the step from accurate vision to mild short-sightedness![14] It makes no sense, Robinson argues, to posit wildly divergent analyses of illusions and veridical perceptions when they have so much in common. The continuities, especially in borderline cases, compel a commitment to at least a few common elements.[15]

Robinson's point spells trouble for Disjunctivism. Consider the case of seeing a certain cup. The Problem of Secondary Qualities says that, when you direct your eyes toward the cup, you see something coloured. But the cup is not coloured. So the colour must belong to some idea, which is the immediate object of perception. Perhaps most Disjunctivists would avoid speculating about the nature of the immediate perceptual object in the illusory case, but bare-bones Disjunctivism, which accepts OP and NPT, agrees that you immediately perceive the colour and that the colour isn't possessed by a physical body. So the immediate object of perception, namely whatever possesses the secondary quality, belongs to an idea. However, rejecting SP, Disjunctivism refuses to extend this analysis to perceptions of the cup's shape and magnitude, which belong to the physical cup and not an idea. There is no mediating mental object in those cases. So simultaneously you see both the physical cup, with its shape and magnitude, and some mediating idea, with its colour.

On balance, the bare-bones Disjunctivist's analysis seems peculiar, placing humankind in dual perceptual worlds of ideas and physical bodies. Although this is coherent, it is not intuitive. To see this, try an actual case. Fetch yourself a mug of fresh-brewed coffee or tea, and consider your perceptions of it. Sight first. The size and shape of your vessel are physical. In observing them, you make cognitive contact with the mug and the physical world outside your mind. But the colours 'of the mug and coffee' are not really of the mug or coffee. They belong to ideas. So overall your visual state is somewhat mixed, incorporating perceptions of both physical bodies with certain shapes and magnitudes as well as coloured ideas. Lifting the mug, you feel its heft, solidity and surface texture, which reside in the physical object. But the warmth that seems to permeate your fingers is mental, again belonging to an idea rather than the mug or your hand. Whether the colour and warmth belong to the same idea, I have no way of knowing. Taking a sip, you may detect the drink's viscosity, which belongs to the physical liquid flowing into your mouth. But the delicious taste and smell inhabit ideas and not the coffee. On this version of Disjunctivism, mental and physical objects mix in concert to produce a unified phenomenology. This is a peculiar world indeed, not the world of common sense.

No wonder Robinson finds the Spreading Principle so appealing. With bare-bones Disjunctivism, you might be looking at a stick and a glass of water and seeing them immediately. But the moment you dip the stick into the water, a bent idea begins mediating between you and the stick. You can't see the stick directly, unless you again remove it from the water. But pull it up, end the illusion and – surprise! You're back in direct cognitive contact with the stick. Does it not seem farfetched to suppose so many physical and mental objects of perception floating about, perhaps even popping in and out of existence?

Perhaps other varieties of Disjunctivism sport a higher degree of plausibility than the one entertained here. Recall that this bare-bones Disjunctivism rejects SP while accepting OC (that humans perceive secondary qualities) and NPT (that physical entities do not possess secondary qualities). But this turns out to involve a bizarre perceptual world, part physical and part mental, which doesn't gain much dialectical territory in the battle for common sense. On the face of it, Idealism offers more plausibility, or at least simplicity. So sticking with the Spreading Principle and holding out in hopes of finding a weakness in the Observation Claim or the Non-Physicality Thesis seems prudent for now.

An Adverbialist Objection to the Spreading Principle

Although not a Disjunctivist, the philosopher closest to the Disjunctivist approach is perhaps Mazviita Chirimuuta. His Fregean Complex Representational Theory doesn't fit neatly into the Idealist, Indirect Realist or Direct Realist camps, but Chirimuuta certainly wants to flag secondary qualities as exceptional.[16] On his Fregean Representational Theory, colours and other secondary qualities are relations or 'modes of presentation' between perceiving subjects and perceptual objects, not intrinsic properties of perceived physical objects. This distinguishes them from primary qualities, which really do inhabit physical objects and get perceived, along with the objects themselves, as a complex, colour-modified whole. Primary qualities, then, are objects. Secondary qualities are mere modes.

Of course, Chirimuuta's view lacks common sense appeal for some of the same reasons as Idealism and Indirect Realism. If colours are perceptual modes rather than objects of perception, they sure don't appear to be. What's more interesting, however, is the fact that Chirimuuta argues for the Fregean theory over a 'simple', 'Russellian' view by appealing to the way that secondary qualities get caught up with primary qualities in a complex representational whole.[17] He understands that the coffee's colour, smell and warmth affect the senses together with its viscosity, weight and size. This is an apt observation, but it raises a question about why secondary qualities deserve the special treatment Chirimuuta proposes. Why posit a perceptual world of modes in addition to a world of objects, especially when the distinction isn't perceptually apparent? As with the bare-bones Disjunctivist, the reasoning behind the Spreading Principle presses for consistency, asking why not objects only or modes only. It's no good making two realities when one will do. And it is doubtful that the Fregean theory can resolve such tension.

When I drink from my coffee mug, its colour seems a part of the mug, just as its shape is. I enjoy the coffee's flavour, warmth and texture as a collective whole. If mental and physical objects of perception are mixed as bare-bones Disjunctivism and Chirimuuta's theory suggest, then Robinson is correct in his criticism that the sharp break between perceptions of physical bodies and perceptions of ideas or modes of perception fails to reflect the phenomenal, causal and physiological continuities between veridical perceptions and slight illusions. This is not the route to upholding common sense. How strange to think that I might taste an idea or mode while drinking a physical liquid!

Although Behaviourism seemed to offer a response to the Problem of Secondary Qualities by rejecting SP, its promises went unfulfilled. Even if its attempt to differentiate secondary qualities from other perceivable properties were convincing, it created additional problems since it may not be a species of Direct Realism. Disjunctivism, including Fregean Adverbialism, creates problems by positing radical differences between physical objects on the one hand and ideas or modes on the other, even when primary and secondary qualities seem integrated into a single physical, perceived world.

Finally, even if one found good reasons to reject SP, it might work against common sense as much as in its favour, because, as Chapter 1 points out, common sense understands secondary qualities as properties of physical objects. So, if one denied SP, the systematic illusoriness of secondary qualities would still undermine common sense. And for this book, common sense is a primary motive for endorsing Direct Realism. Direct Realism derives its basic appeal by affirming that the world is mostly as it seems.

Defending Direct Realism by admitting that ever-present second-ary qualities are not as they seem constitutes a loss of dialectical ground. So it's best to accept SP. If a convincing argument against it ever arises, preferably one that supports Direct Realism, then perhaps the Problem of Secondary Qualities will be exposed as weaker than it seems today. Today it seems strong.

The Secondary Quality Observation Claim

Careful examination of the first premise of the Problem of Secondary Qualities, the Spreading Principle, didn't reveal much weakness. SP has the support of philosophical tradition as well as intuitive appeal. So consider the next premise, the Observation Claim (OC) – that some immediate objects of perception possess secondary qualities. This is so obvious that it may not need defending. Sense experience itself testifies to OC. Secondary qualities seem to be everywhere. And common sense Direct Realism would like to say that things, for the most part, are as they seem. Even Indirect Realists find this thought appealing. Robinson, for example, motivates his theory with a Phenomenal Principle: 'If there sensibly appears to a subject to be something which possesses a particular sensible quality then there is something of which the subject is aware which does possess that quality'.[18] OC would be difficult to deny given its broad appeal. However, this section ventures two possible objections – gambits really – before addressing the Non-Physicality Thesis (NPT).

The Grand Illusion Objection

Consider the view that secondary qualities do not exist. That is, although immediate objects of perception seem to possess colours, smells and tastes, they don't in fact. Secondary qualities don't exist, either mentally or physically. Despite appearances, what if it's all a grand illusion? Then OC is false, and no matter how things seem, Direct Realism is true. Some physical objects are immediate objects of perception, regardless of what anyone thinks she feels or hears.

The Grand Illusion objection is right that things are not always as they seem. Robinson's Phenomenal Principle is too strong. It makes perceptual beliefs about immediate objects virtually infallible – since immediate objects must be as they 'sensibly appear'. This is why Robinson needs his Sense-Datum Theory. Physical objects, he thinks, fail to satisfy the Phenomenal Principle, so there must be ideas about which perceptions give infallible beliefs. Recall the problems associated with this approach, outlined in Chapter 1. Idealism and Indirect Realism fail to conform to the Phenomenal Principle because at least some ideas, if they exist at all, appear to be physical rather than mental. That is, ideas aren't immune to illusion.

Michael Huemer suggests a more tempered approach to Robinson's Phenomenal Principle, saying that although sense experiences offer significant justification for perceptual beliefs, those beliefs are defeasible.[19] Additional evidence might make it reasonable to reject a perceptual belief. Huemer's Direct Realism acknowledges the reality of illusions but remains optimistic about the veridicality of perception-produced beliefs in general. For example, he concedes that the Müller-Lyer illusion (Fig. 3.1) is a bona fide illusion, but he denies that it undermines the faculty of perception altogether. About that case he says,

> The [centre line] appears to be longer than the [top line], so other things being equal, you would be reasonable in thinking that the [centre] line is longer. However, you can get out a ruler and measure these lines. If you do this, you will find them to be of the same length. At this point it would be reasonable for you to revise the initial belief and conclude that the lines are really of the same length.[20]

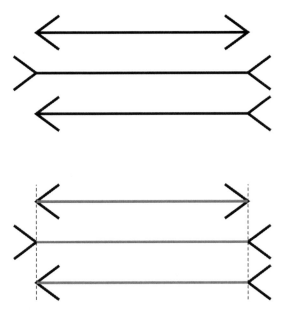

Figure 3.1 The Müller-Lyer illusion

Naive beliefs about the relative lengths of these lines may turn out to be inaccurate. Vision alone won't cut it, so other epistemic faculties and instrumentation count for a lot. Even if perceptual abilities are impressive, they can fail.

The Grand Illusion objection to OC says that secondary qualities do not exist either as physical or mental properties. It dismisses secondary quality perceptions as false testimony from imperfect perceptual faculties, as in the Müller-Lyer illusion. One line looks longer to you, but nothing is in fact longer. Likewise, you seem to see secondary qualities. You have secondary quality sensations. But according to the Grand Illusion objection, there are no secondary qualities or perceptions of them. You can dream of unicorns, but you can't see them, for there are none to be seen. On the Grand Illusion objection to OC, perception-based beliefs in secondary qualities are just epistemic misfirings.

Such a move is not without risks. Problems in perception tend to ramify, as the Spreading Principle illustrates. Consider that for every illusion or supposed illusion, there might be a corresponding spreading principle. Each illusion raises a question about its implications for perception generally. Is it a special case? Or does it mean that perceptual beliefs somehow fail universally? The Müller-Lyer, for example, may be a bona fide illusion, but even if it is, why should that suggest that you can't see the width of a sidewalk or the shape of a street sign? It is no surprise that human visual and neural systems have limits, but limits are not necessarily debilitating. The Müller-Lyer strikes Huemer and others, including me, as mostly benign. But are there cancerous illusions that threaten to spread to perception generally, thus threatening Direct Realism and common sense? Consider a world in which trees are illusory or people or physical motion or the passage of time. Any of these might prompt you to reconsider your commitment to perceptual belief altogether. Theories of perception can tolerate imperfections, but they have their breaking points.

It is difficult to say exactly which cases threaten perception's overall epistemic status. But secondary qualities feature heavily in everyday perceptions, and perceivers face a near-irresistible urge to believe in them. In this respect, secondary qualities fit alongside trees and motion better than with Müller-Lyer drawings. If they turn out to be illusory, then humanity has been fantastically duped by its own epistemic equipment. Direct Realism can't go here. Moreover, Idealists and Indirect Realists should reject the Grand

Illusion objection to OC as well, because they have a high view of perception's epistemic value too. Treating secondary qualities as a grand illusion wreaks havoc on the whole discussion.

The No-Qualities-No-Illusion Objection

The approach in the last section suggested denying altogether that there are secondary qualities while admitting that human perceivers seem to perceive them. It didn't work so well. Secondary qualities inhabit too much of the perceptual realm to fail benignly. But this is not the only way to reject the Observation Claim (OC). Suppose that, as with the Grand Illusion, there are no secondary qualities. But perhaps no one perceives them or even seems to perceive them. Secondary qualities aren't illusions, just misleading circumlocutions or relics from a dubious and outmoded 'folk psychology'. Is this a plausible means for rejecting OC and so defeating the Problem of Secondary Qualities? Almost certainly not, because such an approach fails to cohere with common sense.

One way to claim both that you do not perceive or seem to perceive secondary qualities is eliminative materialism about the mind. Paul Churchland, for example, predicts that a completed neuroscience will make traditional 'folk psychology' obsolete.[21] Concepts of desires, beliefs, and perhaps even concepts will give way to a new, scientific view of the mind. Churchland suggests that in light of science's conclusions, philosophers and others will jettison conceptual schemes involving propositional attitudes of the mind, including beliefs, memories, perceptions, hallucinations, and illusions.[22] And without perceptions and illusions, how can you say that you perceive or misperceive or have illusions of secondary qualities? How can anyone say that objects of perception possess secondary qualities? The Problem of Secondary Qualities becomes mute. There are no perceptions. So someone like Churchland could easily reject OC and the supposed evidence for it. Immediate objects of perception don't possess secondary qualities, nor do they seem to.

The problem here is that two thousand years of philosophical tradition say otherwise. In Scholastic treatments of perceptual objects, secondary qualities often serve as paradigm cases. Aquinas, Lutrea (the writer of a respected commentary on Aristotle's *De Anima*), Usingen (Martin Luther's teacher at Erfurt), Buridan and Ockham all treat the secondary qualities as physical and often use colour as a perceptual paradigm case.[23] But how could such a tradition arise unless perceivers seem to perceive secondary qualities? This conviction of Scholastic philosophy has continued to the modern era without resistance, as one of the few points where philosophers have near-consensus. As C. D. Broad puts it, 'What I am immediately aware of when I look at a penny stamp is as indubitably red as it is indubitably more or less square'.[24]

Furthermore, the intuitiveness of OC finds its way into everyday human language, a significant indication of common sense appeal, according to Thomas Reid.[25] If there are no such things and don't seem to be such things as colours, sounds or tastes, then 'the colour of the tulip', 'the sound of the bell' and 'the taste of the apple pie' need reinterpretation. What could these phrases mean if not that human subjects can perceive secondary qualities of physical objects? Are they mere colloquialisms? If so, then they are fantastically widespread and horribly misunderstood.

Eliminative materialism, with its metaphysically revisionary elimination of folk psychology, explicitly places itself in opposition to common sense. While common sense does not always track the opinions of the masses, it does endorse natural inclinations to trust epistemic faculties, like perception, memory and reason. Jettisoning the whole notion of perception, then, would be a bizarre sidestep, not a way to save Direct Realism from the Problem of Secondary Qualities.

Perhaps someone will find other ways to attack OC that are more friendly to common sense and Direct Realism. For now, the second premise of the Problem of Secondary Qualities remains convincing. Indeed, OC may be the most firmly rooted commitment in this discussion. Direct Realism and its rivals, Indirect Realism and Idealism, all say that secondary qualities are among the objects of human perceptions.

Having found no suitable weaknesses in SP or OC, the next step is to address the Problem of Secondary Qualities' final premise, the Non-Physicality Thesis.

Are Secondary Qualities Physical?

The page has come to address NPT, that if an object is physical, then it does not possess secondary qualities. This premise, I contend, is the weak point in the Problem of Secondary Qualities. Part II explains Thomas Reid's acceptance of secondary qualities as physical properties. But before moving on to Reid, it is only fair to explain the Problem of Secondary Qualities' case for NPT. Interestingly, Robinson's defence of NPT begins in the High Renaissance, with Galileo Galilei.[26]

Galileo offers an early commitment to secondary qualities as subjective. He writes that the sensations associated with secondary quality perceptions result from 'other' and 'real' qualities. Tastes, for example, depend on the 'various shapes, numbers, and speeds' of the particles that contact the tongue. Smell sensations come about likewise via the nose. Sounds are heard at the 'ruffling' of the air. And although he offers no comment on colour, Galileo theorises that heat sensations originate with tiny moving particles that penetrate human bodies. Fully aware of secondary quality sensations, he doesn't see that sensations alone support belief in secondary qualities as physical properties, especially if the causes of secondary quality sensations can arise from other, 'real' properties. Galileo concludes, 'My mind does not feel compelled to bring them in as necessary accompaniments [to physical substances]'.[27]

To support his view, Galileo tries a thought experiment. Consider a possible world with the same physical objects and properties as the actual world, but with the exception that there are no secondary qualities possessed by physical objects. In that world, says Galileo, your sensations would be the same as in this world, because primary and other physical properties completely account for the physical causes of sensations. That is, the possible world where physical objects do not possess secondary qualities is phenomenologically indistinguishable from this world. Therefore, Galileo reasons, secondary quality sensations do not count as evidence for secondary qualities as physical properties, which leaves human perceivers no reason to think that secondary qualities are in fact possessed by physical objects.

What makes Galileo think that secondary qualities would make no difference to one's phenomenology? He and other early modern natural philosophers are Atomists, Mechanical Philosophers, neo-Epicureans. Having jettisoned Scholastic formal causes, Galileo and his friends see the universe as consisting of minute atoms, governed solely by collisions and geometric features.[28] Since mechanical interactions account for the whole of physical causation, there is no causal role for secondary qualities. Robert Adams summarises,

> The only properties of bodies directly involved in such interactions are geometrical properties (size, shape, position), motions (and rest, which is lack of motion), and solidity (the fact that two bodies cannot occupy the same space). These are the qualities

that came to be called 'primary'. The other sensible qualities are assigned no direct causal role in this scheme of explanation.[29]

If secondary qualities play no causal role in the processes that determine your sense experiences, then, even if physical objects happen to possess secondary qualities, these are not the ones that you perceive, for you have no causal connection to them. Therefore, you have no good reason to think that the secondary qualities you perceive are physical and objective. Adams again:

> When I see a round, red apple, the geometrical property of roundness that is immediately present to my mind may on this view also be present in the apple, playing a causal role in the effect the apple has on my eye, and ultimately on my mind. But the quality of redness that is immediately present to my mind seems to be entirely different from, and additional to, any geometrical property or motion. As such, if it is present outside the mind, in the apple, it cannot have any effect on my eye, or on any other body, according to the exclusively mechanical theory of explanation.[30]

The early moderns reason that the causal connection of mind and world is limited by sense organs. Perceivable secondary qualities only make sense as ideas, illusory but only in so far as they seem physical and in fact are not.

One might be inclined to try identifying secondary qualities with or reducing them to properties accepted by Atomists – size, shape and motion. Then, even if individual atoms lacked secondary qualities, combinations or groups of particles might possess them collectively. The following chapters pursue a solution along these lines, but the early moderns, like many contemporary thinkers, are against it. Part III explores and challenges some of the reasons why the identity option does not find much sympathy among philosophers and scientists.

Do today's scientists and philosophers also hold Galileo's view towards secondary qualities? A cursory glance at the indices of several chemistry and physics textbooks suggests that, with the exception of heat, the traditional secondary qualities are not subjects of interest for contemporary natural scientists, at least not by their traditional names.[31] And it is not difficult to find a few who testify to NPT, especially with regard to colour. For example, Stanford University biologists Robert Ornstein and Richard Thompson:

'Color' as such does not exist in the world; it exists only in the eye and brain of the beholder. Objects reflect many wavelengths of light, but these light waves have no color. Animals developed color vision as a way of telling the difference between various wavelengths of light. The eye converts different ranges of wavelengths into colors, and it does this in a very simple way.[32]

And Leda Cosmides and John Tooby, cofounders of the Center for Evolutionary Psychology at the University of California-Santa Barbara:

Just as common sense is the faculty that tells us that the world is flat, so too it tells us many other things that are equally unreliable. It tells us, for example, that color is out there in the world, an independent property of objects we live among. But scientific investigations have led us, logical step by logical step, to escape our fantastically insistent, inelastic intuitions. As a result, we know now that color is not already out there, an inherent attribute of object. We know this because we sometimes see physically identical objects or spectral arrays as having different colors . . . Far from being a physical property of objects, color is a mental property – a useful invention that specialized circuitry computes in our minds and then 'projects onto' our percepts of physically colorless objects.[33]

Despite their philosophical clumsiness (saying that the contents of eyes and brains do not exist in the world), these scientists' position is accessible. According to them, scientists know the physical causes of colour appearances, and those causes are not colour properties in physical objects or in reflected light. It only makes sense to speak of colours when considering the visual or neurological physiology of some organism. Colours, and by extension other secondary qualities, can't be physical and outside a body. Hence, the Non-Physicality Thesis. Notice that Cosmides and Tooby base their claim on the lack of correlation between sensations and scientifically familiar physical properties, a theme that features again in Part III.

NPT receives some support from the testimony of scientists. Tough-minded analytic philosophers prefer to accept the findings of science, because philosophers and scientists have the same fundamental goal of understanding human beings and the world they inhabit. Moreover, the systematic and experimental rigor of the natural sciences often yields a greater degree of certainty than is typical of philosophy. W. V. Quine expresses it well, saying, 'Our ontology is determined once we have fixed upon the over-all conceptual scheme which is to accommodate science in the broadest sense'.[34] However, the Problem of Secondary Qualities is a special case. Rejecting NPT could save Direct Realism. So it's worth considering the evidence for NPT to see whether its supporters correctly treat it as an invincible result of rigorous scientific investigation rather than a lingering anti-Scholastic attitude. So start with a contemporary argument for NPT.

For one formidable argument that physical objects do not possess secondary qualities, consider Frank Jackson's case. Jackson concerns himself with colour in particular, warning against regarding colour as possessed by physical objects unless it is a scientific property. 'Scientific property' is a term of art: 'A scientific property is a property appealed to by current science in explaining the causal effect of one material thing on another material thing, or a logical consequence of such a property or properties.'[35] As examples, Jackson names mass and charge, qualities involved in physical causation. Notice that on this definition scientific properties include those which may be scientifically identified as combinations of other scientific properties – like torque or electric flux – so some scientific properties may be reducible to or identified with other scientific properties.

Jackson begins, 'Our reason for believing that material things are coloured is certain of the perceptual experiences we have'. He means to recognise a unique role for sensations in forming beliefs to the effect that certain physical things are coloured. For the sake of formality, one might phrase it, 'Unless sensations offer good reasons to believe that physical objects are coloured, there is no reason to think that those objects are in fact coloured.' For Jackson as for Galileo, sensation offers the only possible epistemic link to physical secondary qualities.

Next Jackson considers the physical causes of sensations, arguing that scientific properties sufficiently account for the brain events that cause sensations. He says,

> Now it is known in broad outline how a material thing causes the brain events relevant to sensory experience. For those experiences particularly relevant to our perception of colour, the process involves the action of light reflected from the object to the eye. And the role of the object is essentially that of modifying the wave-length

composition of the light, and the properties of the object which effect this modification are scientific ones like the texture and the molecular structure of its surface. . . . we do not yet know the (operative) necessary and sufficient conditions in full detail, but we are far enough along the road to knowing them to be able to predict with fair confidence that they will not require us to invoke properties over and above those countenanced by current science.[36]

In only a few lines, Jackson has added two more premises. First, the immediate causes of sensations are certain brain events. Second, causal effects on brains depend solely on scientific properties of physical objects. No rational person wants to attribute secondary qualities to physical objects without good reason – in this case, the evidence of sensation. But if the evidentiary potency of sensations depends on causal connections between brains and physical properties, then sensations can only evidence physically causal properties. In sum, if colour is a non-scientific property, then sensations can provide no reason to believe that physical objects are coloured. With sensations, the evidence for colour becomes threadbare, and common sense faces charges of irrationality. For beliefs in physical colours to pass muster, colour must be a scientific property.

According to Jackson, whether colour and, by extension, other secondary qualities are possessed by physical objects depends on whether they can feature in causal-scientific explanations of the sort offered by current scientists. Jackson concludes that they can't. Scientific explanations, he says, make no appeal to colour: 'There is not one causal law in which "is red", "is blue", and so on appear.'[37] Position and shape help make scientific predictions. Colours do not. Rather than secondary qualities, science uses 'hypothetical movements of hypothetical parts', which have the causal and predictive powers necessary for scientific theories.[38] Not only is colour not mentioned in scientific descriptions of the physical process leading up to human perceptions, it is not even a prospective property for such a role. Colours are not scientific properties, so there is no reason to regard them as physical. And Jackson could easily say the same of other secondary qualities.

Jackson's argument offers something beyond Robinson's version of the Problem of Secondary Qualities, a reason for NPT. No physical process resulting in a colour perception involves a physical colour. Likewise for the other secondary qualities. Given contemporary knowledge of the human perceptual-nervous system, who could argue with Jackson on these points? The causal mechanism between external objects and perceiver's mind is physical, or scientific. Therefore, if secondary qualities are possessed by physical objects and perceivers learn about them through sense, then secondary qualities must engage physical-causal mechanisms. This means challenging Jackson's claim that colours are not scientific properties and ultimately NPT. The remainder of this book attempts this by identifying secondary qualities with known physical-scientific properties.

Conclusion

Direct Realism is appealing because it harmonises human epistemic faculties, yields theoretical justification for common sense perception-based beliefs, and is more intuitive and less problematic than its idea-driven alternatives. But the Problem of Secondary Qualities presents a serious philosophical challenge to Direct Realism.

The Problem of Secondary Qualities pits epistemic optimism, sensations and scientific findings against one another. To defend Direct Realism, I propose an attack on the scientifically motivated claim that physical objects do not possess secondary qualities, the Non-Physicality Thesis. But NPT is a substantiated claim. For over 400 years, scientists and philosophers have insisted that there is no reason to regard secondary qualities as inhabiting the external physical world, especially since they cannot be identified with scientific properties. Secondary qualities seem causally superfluous, experimentally non-predictive, and theoretically impotent.

I propose to say, contrary to Jackson, that secondary qualities are scientific or, more specifically, identifiable in terms of scientific properties. They are implicitly referenced in scientific explanations even if the references are not obvious. Science can treat a certain property while sense perception reveals different aspects of the same property, and scientists may not recognise it as a single property. If so, then the Problem of Secondary Qualities disappears, since secondary qualities are possessed by physical objects and NPT is false.

Critics of Direct Realism claim that the proposed programme to identify secondary qualities with physical-scientific properties is a dead end. Part II introduces Thomas Reid's theory of perception and his account of primary and secondary qualities, showing how his theory of secondary qualities involves their identification with scientific properties. Part III consists of objections and replies.

Notes

1. Berkeley, *Three Dialogues*, p. 38.
2. Berkeley, *Treatise Concerning the Principles of Human Knowledge*, 1.10/106.
3. Galileo, *The Assayer*, pp. 276–7.
4. Hume, *Treatise on Human Nature*, 1.4.4.6–10/150–1.
5. Dancy, 'Introduction', p. 31.
6. Foster, *The Nature of Perception*, p. 65.
7. Foster, *The Nature of Perception*, p. 65.
8. Pitcher, *A Theory of Perception*, p. 222.
9. Pitcher, *A Theory of Perception*, p. 228.
10. Pitcher, *A Theory of Perception*, pp. 199–202.
11. Elliot and Niesta, 'Romantic Red: Red Enhances Men's Attraction to Women'.
12. Berkeley, *Three Dialogues*, p. 25.
13. Soteriou, 'The Disjunctive Theory of Perception'.
14. Robinson, *Perception*, p. 57.
15. Robinson, *Perception*, pp. 56–8.
16. Chirimuuta, *Outside Color*, pp. 177–8.
17. Chirimuuta, *Outside Color*, pp. 167–70.
18. Robinson, *Perception*, p. 32; Fish, *Philosophy of Perception*, p. 6.
19. Huemer, *Skepticism and the Veil of Perception*, pp. 98–103.
20. Huemer, *Skepticism and the Veil of Perception*, p. 101.
21. Churchland, 'Eliminative Materialism and the Propositional Attitudes', pp. 67–90.
22. Churchland, 'Eliminative Materialism and the Propositional Attitudes', p. 87.
23. Karkkainen, 'Objects of Sense Perception in Late Medieval Erfurtian Nominalism'; Ivery, 'The Ontological Entailments of Averroes' Understanding of Perception'.
24. Broad, 'The Theory of Sensa', p. 125.
25. For example, see Reid, *Essays on the Intellectual Powers of Man*, 1.1/21, which notes that certain operations of the mind are expressed, in every language, by active verbs and so suggests that 'the natural judgment of mankind' considers the mind to be active.
26. Robinson, *Perception*, pp. 59–60.
27. Galileo, *The Assayer*, pp. 274–7.
28. Chalmers, 'Atomism from the 17th to the 20th Century'.
29. Adams, 'Editor's Introduction', pp. xiv.

30. Adams, 'Editor's Introduction', pp. xiv–xv.
31. One notable exception, Silberberg, *Chemistry*, pp. 1026–7, includes a confusing three-paragraph section titled 'What is Color?' The question that forms the title goes largely unanswered, and the language is erratic. Silberberg sometimes claims that colours are properties of things and other times that they are types of light. He sometimes says that an object has a colour and other times that it merely appears to have it.
32. Ornstein and Thompson, *The Amazing Brain*, p. 55.
33. Cosmides and Tooby, 'Forward', pp. xi; quoted in Tye, *Consciousness, Color, and Content*, p. 145.
34. Quine, 'On What There Is', p. 36.
35. Jackson, *Perception*, p. 122.
36. Jackson, *Perception*, pp. 124–5.
37. Jackson, *Perception*, p. 127.
38. Broad, 'The Theory of Sensa', pp. 126–7.

PART II

How Thomas Reid Solves the Problem

Part I introduced the Problem of Secondary Qualities as an important challenge to Direct Realism and to common sense. The Problem of Secondary Qualities rests on three premises: the Secondary Quality Spreading Principle (SP), the Secondary Quality Observation Claim (OC) and the Secondary Quality Non-Physicality Thesis (NPT).

Perhaps surprisingly, SP and OC proved unrelenting upon cross-examination, which raised the question of whether NPT might harbour dialectical weaknesses. Having reviewed the case for NPT, Part I ended with a call for a theory of secondary qualities on which they are possessed by physical objects and identical to physical, scientific properties. To do this, Part II suggests just such an account, derived from the writings of Scottish Common Sense philosopher Thomas Reid.

Reid's writings on perception are radically polemical. He targets the Way of Ideas, a philosophical tradition that he traces from his contemporaries, including David Hume, all the way back to Pythagoras. According to Reid, it holds that all immediate objects of perception, indeed all immediate objects of thought, fall within the realm of the mental. They are ideas. And what consequences arise from such a view? Hume, Reid thinks, has it right: external world scepticism. Can human perceivers learn anything about the physical world, even whether it exists? Finding this conclusion unacceptable, Reid develops his *Inquiry into the Human Mind on the Principles of Common Sense* as an alternative to the received view. To this end, he presents and defends his account of primary and secondary qualities as objects of sense perception, and as properties of physical objects. Sensations form a gateway to the physical world.

As it happens, the proper handling of Reid's writings about primary and secondary qualities and the exact details of Reid's philosophy of perception remain controversial among Reid scholars. How does he theorise that minds gain knowledge of the physical world by means of sensations and sense organs? How do the theoretical elements from Reid's account of perception provide the conceptual building blocks for his view of primary and secondary qualities? How does his theory undermine NPT? And how should heated questions about interpreting Reid's work find resolutions?

The following chapters present Reid's understanding of primary and secondary qualities in the context of his broader philosophy of perception. Chapter 4 establishes the conceptual landscape of Reid's analysis of perception, including relations among perception, belief, conception and sensation. To grasp his primary/secondary distinction, it considers the conceptual nature of perception, the contingent connection between sensations and perceptions, and the role of sensations as natural signs. The same chapter presents a new interpretation of Reid on primary and secondary qualities, supported by Reid's general claims about the primary/secondary quality distinction as well as his commitments regarding individual primary and secondary quality species. Ultimately, Reid identifies secondary qualities with physical, scientific properties, thereby rejecting NPT and answering the Problem of Secondary Qualities. Details of the rejection and answer receive further treatment in Chapter 5. Chapter 6 addresses interpretive minutiae for the sakes of honouring Reid, engaging Reid scholars and entertaining the exegetically curious.

4

Primary and Secondary Qualities in Reid's Theory of Perception

According to Reid, perception is an operation of the mind by which one becomes aware of external objects. He names three main features that collectively distinguish perception from other mental operations, like imagination, memory and judgement: (1) a conception of the object perceived, (2) an irresistible conviction and belief of its present existence, and (3) a cognitive immediacy regarding the conviction and belief. That is, a perception consists of two components, a conception and a belief, arising independent of reasoning, argument or other mental deliberation.[1] The first feature makes perception intentional or object-taking. The second separates it from operations pertaining to non-existent or past objects, like those involved in imagination and memory. The third makes perception independent of inference. For Reid, perception means drawing on some characterisation of the perceptual object and gaining confidence that the object exists without constructing syllogisms or analysing data. In addition to these three, Reid emphasises the causal chain connecting a perceived object to someone's perception of it.[2] Perception grants you the raw ability to know things and properties outside yourself.

On Reid's scheme, perception deploys conceptions with regard to objects, and in virtue of these conceptions, perception is intentional, having the ability to take objects. A perceptual conception is *of* the perceived object, therefore the perception is *of* the object. The intentionality of the perception, its object-taking, arises in virtue of the intentionality of the conception. You might express a conception in verbal-descriptive terms – for example, 'the computer on the desk', 'the dog on the neighbour's porch' or even 'that thing left of the tree'. Thus, conception's contribution to the overall perception includes two aspects, an object-taking and a mental categorisation or labelling of that object. It involves thinking *of* an object *as* something.

It may seem odd that Reid considers belief in the perceived object essential for perception. Conception alone, like imagination or simple apprehension, is not enough. Conception, on Reid's view, carries no judgement, no conviction, about the existence of any object.[3] So Reid adds belief, a mental act of affirmation or denial, to distinguish perception from mere imaginings or musings. Unlike conception, it is propositionally expressed if at all.[4] While you might verbally represent a conception of a certain computer as 'the computer on the first desk', any adequate expression of a belief in that computer requires a full statement, like 'There is a computer on the first desk'. Reid claims that, in speech, belief always finds its expression via propositions and that, without belief, there would be no reason for the affirmation or denial of propositions in language.[5]

Beyond these two features, Reid treats belief as a conceptual simple, which admits of no 'logical definition' or philosophical analysis. It is an intentional state, involving conception but admitting of degree, depending on the subject's certainty, 'from the slightest suspicion to the fullest assurance'.[6] That is, whereas conceptions may be more or less direct or more or less clear, beliefs can be more or less certain. Along this sliding scale of assurance, Reid generally places perception at the top, as a bearer of certainty, noting its status as a gold standard for evidence in legal proceedings.[7] Reid offers the examples of a star emerging at twilight and a distant ship making its way over the horizon:

> In perception we not only have a notion more or less distinct of the object perceived, but also an irresistible conviction and belief in its existence. This is always the case when we are certain that we perceive it. There may be a perception so faint and indistinct, as to leave us in doubt whether we perceive the object or not. Thus, when a star begins to twinkle as the light of the sun withdraws, one may, for a short time, think he sees it without being certain, until the perception acquires some strength and steadiness. When a ship just begins to appear in the utmost verge of the horizon, we may at first be dubious whether we perceive it or not: But when the perception is in any degree clear and steady, there remains no doubt of its reality; and when the reality of the perception is ascertained, the existence of the object perceived can no longer be doubted.[8]

Some interpreters take this passage as evidence for perception without belief.[9] True, Reid muddies his language a bit here, but the first sentence makes his point clear – until the belief is beyond doubt, it does not count as a perception. The last line is the same. Once the perception is present, all doubts are expelled. Perception requires not merely belief but irresistible belief.

For Reid, as for many contemporary philosophers, 'perception' is a success term.[10] There is no perception unless the object and the mental operation connect appropriately. This requires some qualification of Reid's theory beyond the three earlier-mentioned features of conception, belief and cognitive immediacy, since by themselves they allow any immediate (non-reasoned) conception and belief, including illusory and hallucinatory cases, to count as perceptions. An interpreter could easily misread Reid on this issue, since his examples almost always use successful cases that effectively dodge the question of whether perceptions essentially take objects. When discussing a phantom limb case, Reid even says that the amputee perceives a non-existent toe. However, within the same passage, he qualifies the mental operation as a 'seeming' perception.[11] In another place, he explicitly prohibits perceptions of non-existent objects as logical impossibilities.[12]

Ultimately, Reid identifies perception as a knowledge-generating mental operation apart from other operations, like inference and abstraction. Moreover, its causal requirements indicate that perceptual knowledge is knowledge of particular things and their qualities, not of universals or imaginary or abstract objects. Both knowledge-generation and intentionality of perception distinguish it from mere sensation.

For pursuing a solution to the Problem of Secondary Qualities, the remainder of this chapter focuses on the features of Reid's thought that are most useful, making only slight alterations when necessary. It is worth mentioning, however, an acute problem with Reid's account of perception, the difficulty concerning time lags in causal chains leading up to perceptions. Reid distinguishes perception from memory by saying that its object must be contemporaneous with its being perceived.[13] This won't work if you can see stars that have been extinguished long before their light reaches Earth. Causation isn't instantaneous, even in sense perception. Still, the main tenets of Reid's theory remain highly plausible. Perhaps the contemporaneity problem can be overcome with the right account of the causal connection between perceived object and perceiving subject, since the causation of perception certainly differs from that of memory. On balance, Reid's account seems reparable, so please don't dwell on this shortcoming. I point it out merely for the sake of disclosure.

Three Approaches to the Content of Perception

J. Todd Buras discusses object-taking and mental labelling aspects of Reidian perceptual conception as types of perceptual contents, 'presentational' and 'referential' contents.[14] Both come about in virtue of the essential role that conception plays in Reid's theory, not by way of sensations. As already mentioned, conceptions have a descriptive aspect, perhaps best expressed in language although not linguistic entities themselves. And in cases of perceptual conception, they also take objects. For Buras, the presentational content of conception (and so perception) is the way in which you comprehend the perceptual object. Following Nicolas Wolterstoff, he suggests that perceivers communicate this content in sentences like 'The subject perceives of the table that it is hard'. The subject brings the perceived object under the conception hard. In Buras's language, the referential content of a conception (and so perception) is its object. In the previous case of the hard table, the referential content is the table itself or the table's hardness. Formally, Buras proposes that the presentational content of a conception might be given by the value of F in a predicative statement of the form 'Fb', where b is the object of the conception. The referential content is the thing to which the presentational content applies. It is the b in 'Fb'.

Some have thought that making conception an essential part of perception creates problems in accounting for perceptions of ignorant or non-human subjects, or even those with different conceptual backgrounds. If conceptions with presentational contents are required for perceptions, then how, for example, can a primitive tribesman in Borneo see a typewriter?[15] Certainly not by bringing the typewriter under the concept typewriter. N. R. Hanson offers another example:

Consider two microbiologists. They look at a prepared slide; when asked what they see, they may give different answers. One sees in the cell before him a cluster of foreign matter: it is an artifact, a coagulum resulting from inadequate staining techniques. This clot has no more to do with the cell, *in vivo*, than the scars left on it by the archaeologists spade have to do with the original shape of some Grecian urn. The other biologist identifies the clot as a cell organ, a 'Golgi body'. As for techniques, he argues: 'The standard way of detecting a cell organ is by fixing and staining. Why single out this one technique as producing artifacts, while others disclose genuine organs?'[16]

The two scientists are aware of the same object but, as Hanson puts it, 'see different things', as in a Gestalt case.

Reid's understanding of perceptual contents, as described by Buras, handles the case rather well. Both scientists see the cell organ. That is, their perceptions have the same referential contents. But each brings the organ under his own conception. One sees the cell organ as a cell organ, the other as an artefact. Even though the referential contents are the same, the presentational contents differ. Likewise, the typewriter-viewing tribesman may have the typewriter as the object of his perception, the perception's referential content. But the tribesman will think of the typewriter as the large black object in the middle of that field, and not as a writing device. On Reid's theory, the tribesman and an urban technologist have different perceptions, even though they see the same object under the same conditions and with comparable perceptual-physiological apparatuses and sensations. The difference arises in virtue of presentational contents, not referential contents or sensations. As Fred Dretske puts it, 'We see armadillos, galvanometers, cancerous growths, divorcees, and poison ivy without realizing we are seeing any such thing'.[17]

One Reid interpreter, James Van Cleve, suggests that Buras's presentational contents, the mental categorisation or labelling aspect of Reid's account, misconstrues Reid's take on conception. He says that this approach effectively reads Reid's conceptions as Kantian concepts.[18] He proposes an interpretation of conception on which conception can, and in fact always does, take objects, but conception does not necessarily involve the mental labelling of presentational contents. Van Cleve likens Reidian conception to Bertrand Russell's knowledge by acquaintance, which involves cognitive contact with an object but without taking the object to be a specific sort of thing. He wants referential content without presentational content. Van Cleve suggests that this happens because object-taking in perception can use a mental demonstrative or 'pointing out', eliminating the need for presentational contents.

Van Cleve further argues that including both presentational and referential contents in conception leaves too much overlap between conception and belief. Belief already includes presentational content, he complains, so why would Reid bother mentioning it as a factor in perceptual conception when perceptual belief already accounts for everything in the product worth mentioning? It seems redundant.

The textual evidence supporting Van Cleve's reading is scarce. Reid's examples of conception all involve mental labelling as well as the ability to represent the labelling in language, like 'pyramid' or 'obelisk'.[19] This includes conceptions of objects with demonstrative presentational content of the sort referenced by Van Cleve. For example, in original conceptions of secondary qualities, you understand perceivable properties by means of causal relations to particular sensations, as in 'the cause of that lemony taste sensation' or 'the occasion of that rosy smell sensation'.[20] So, contrary to Van Cleve's interpretation, for Reid, when an object is present to one's senses, one's conception of the object does involve presentational content in addition to referential content.

Furthermore, even if belief shares its presentational content with conception, Reid may have broader schematic reasons for distinguishing the two and accounting for both in perception. Perhaps he sees perception as a process developing over time and considers conception and belief as initial and final elements of the process. Second, pruning Reid's notion of conception back to include only referential content does not resolve Van Cleve's worry about redundancy. Perceptual belief has not only presentational but referential content, so perceptual conception with only referential content still yields a redundancy in content, namely, the conception and belief share an object. So the alternative reading bears this supposed problem as well as the one offered here. Finally, the redundancy should not come as a surprise to Reid's readership. After all, as Van Cleve himself notes, in Reid's scheme conception plays a role in every operation of the mind, including belief.[21] If conception plays a role in all belief, then, of course, perceptual belief encompasses or repeats some of what perceptual conception has already contributed. The redundancy is appropriate.

What is not on Reid's list of essential characteristics of perceptions is as important as what is. The key omission is sensation or perceptual experience. What role do sensations play? Reid does not regard them as essential to perception, although sensations happen to trigger perceptions.[22] This causal connection counts only as a contingent fact about human psychology, without bearing on the nature of perception. God might have constructed humankind so as to experience colour appearances when perceiving sounds or even textures.[23] Angels and demons may perceive without sensations at all.[24] Nothing obviously connects sensations to perceptual conceptions and beliefs in a necessary way.

The distinction between sensation and perception features in Reid's attack on the Way of Ideas. Reid sees the conflation of these two as the snare that caught Hume, Berkeley and their predecessors.[25] The first line of Hume's *Treatise* limits mental operations and their objects to two distinct kinds of 'perceptions' – impressions and ideas.[26] Sensations, passions and emotions feature among Hume's impressions, which casts them as species of perception. Reid regards this categorisation as an abuse of language, responding, 'I believe, no English writer ever gave the name of a perception to any passion or emotion'.[27] For Reid, sensation is no perception but a different kind of mental operation altogether. He explains the semantics of his own position,

> Sensation, and the perception of external objects by the senses, though very different in their nature, have commonly been considered as one and the same thing . . .
>
> Thus, I feel a pain; I see a tree: the first denoteth a sensation, the last a perception. The grammatical analysis of both expressions is the same: for both consist of an active verb and an object. But, if we attend to the things signified by these expressions, we shall find, that in the first, the distinction between the act and the object is not real but grammatical; in the second, the distinction is not only grammatical but real.
>
> The form of the expression, I feel pain, might seem to imply, that the feeling is something distinct from the pain felt; yet, in reality, there is no distinction. As thinking a thought is an expression which could signify no more than thinking, so feeling a pain signifies no more than being pained. What we have said of pain is applicable to every other mere sensation.[28]

Mere sensations don't count as perceptions. They are different operations altogether. Most importantly, perceptual objects are external, whereas the object of a sensation is the sensation itself, if sensations have objects.[29] This distinction begins to show where Reid takes Hume and others to have gone wrong. If Hume takes himself to be analysing perception but is in fact limiting his consideration to sensation, then it is no wonder that he doesn't find evidence for an external world or the means to make such a world an object of thought. He ignores perception altogether.

What role, then, do sensations play in perception? For Reid, sensations arise because physical qualities act, mediately or immediately, on the sense organs. Sensations, in turn, trigger perceptions. Perceptions, then, are 'interpretations of sensations', because sensations serve as natural signs for objects of perception.[30] To interpret a sensation, according to Reid, you must form a conception of the thing signified by the sign and a corresponding belief. A heat-type sensation, for example, gives you a conception of and belief in an object's heat, which you rightly understand as a physical quality. Another sensation leads you to perceive the physical quality of hardness.[31] Sensations happen to play this role in human beings, but again only contingently. Reid sees an analogy between sensations as natural signs and the use of words in communication. In the same way that the word 'gold' contingently signifies gold metal, sensations stand for qualities.[32] Thus, sensations are causal in a 'loose and popular', Humean sense of being constantly conjoined to qualities.[33] They participate in the perceptual process. But don't confuse them with perceptions or objects of perception.

You might worry that Reid's theory of perception sounds suspiciously like Indirect Realism rather than Direct Realism. After all, the causal chain between perceived physical property and perception gets mediated by sensations, sense organs and brain activities. But Reid distinguishes immediate objects of a perception and the perception's immediate or mediate causal origins. Why should causal proximity determine a perception's mediacy or immediacy? On Reid's view, an immediate object of perception is one for which the perceiver has a conception and belief but not in virtue of conceiving of and believing in some other object. Reid's mediacy–immediacy chains are intentional, not causal. His theory of perception is direct because of the referential contents of sense perceptions. A perception takes some physical object or quality as an object, not a sensation or preceding causal-physiological event. Smoke may prompt you to think of fire, but thinking of smoke isn't a roundabout way of contemplating fire. Reid has a Direct Realist theory of perceptual objects and a natural sign theory of perceptual causes.

For Reid, sensations work alongside other natural signs. He organises natural signs into three distinct types, which include both mental and physical phenomena. All three types have in common the contingent fact that God or nature has joined sign to thing signified. But Reid sorts them by how human subjects become aware of the connection between sign and thing signified. He differentiates the types as follows:[34]

> Type-1: The natural connection between sign and thing signified is discovered by experience. These connections are the bases for scientific and technical inquiry – mechanics, astronomy, optics, agriculture, gardening, and medicine.

As paradigm cases, he names the constant conjunctions involved in natural laws or deductive-nomological scientific explanations. You have to figure out type-1 connections for yourself. For example, the consistent orientation of a compass needle is a sign of the relative orientation of Earth's magnetic field. In Newton's second law, mass and acceleration serve as signs for the net force on an object, and force signifies the product of mass and acceleration. Smoke is a natural sign for fire. And the sensation experienced when smelling a rose serves as a natural sign for the physical property of the rose called its smell. In all these cases, you discover connections between signs and things signified through experience and scientific research. Type-1 signs include some sensations, like those associated with secondary quality perceptions, but not all.

> Type-2: The natural connection between sign and thing signified is discovered by 'a natural principle', without reasoning or experience.

You have to have some notions of things signified by type-2 signs, but, once you do, nature uses the notions to connect signs to things signified without any real cognitive work. Type-2 signs include facial expressions and musical pieces in so far as they convey emotions. The ability to connect type-2 signs to the things they signify forms the basis of artistic taste, Reid thinks. The key distinction between these signs and type-1 natural signs concerns the way in which one comes to associate them with the things that they signify. Whereas the connections between type-1 signs and the things that they signify are discovered through investigative efforts, the associations between type-2 signs with the things that they signify are granted by nature. Even an infant, Reid says, can understand an angry countenance.

Type-2 natural signs do not play a significant role in understanding Reid's primary/secondary quality distinction. But the third type does.

> Type-3: The natural connection between sign and thing signified is ingrained in the human constitution. 'Though we never before had any notion or conception of the things signified', these signs 'conjure it up, by a natural kind of magic'.[35]

Type-3 signs work even if you have no notion of the thing signified. Nature provides both the notion and the connection, like a built-in translator. All type-3 natural signs are sensations, and nature grants perceivers the ability to properly decipher them. As with type-2 signs, you don't have to discover connections between type-3 signs and things that they signify. But type-3 signs have an additional feature that separates them from the type-2 category. Type-3 signs prompt you to understand the things that they signify as they are in themselves. The paradigmatic example Reid offers is the sensation involved in perceiving hardness. The tactile sensation immediately triggers the conception of and belief in a thing's hardness as the cohesion of its parts with more or less force, even if you have never before considered such a notion. You not only connect the sensation caused by the hardness with the perceived quality, you also learn something about what the hardness is. Moreover, according to Reid, type-3 signs present the only way to begin learning about the natures of physical objects. Without them you would have access only to the natures of things in your own mind, which for Reid means mental operations and states of mind. You would never form direct conceptions of any physical quality, primary or secondary, or find perceptual evidence for the existence of external physical objects.[36]

To summarise, human subjects discover the connections between type-1 natural signs and the things signified by means of experience and discovery, not by nature. The first class of signs does not communicate anything about the natures of the things signified without help from other epistemic avenues. The connections between type-2 natural signs and the things that they signify are given by nature. But you can acquire conceptions of things signified by type-2 signs by means other than the signs. Finally, nature reveals the connections between type-3 signs and the things that they signify, and nature provides conceptions of the things they signify as they are in themselves. Not all natural signs are sensations, and not all sensations are type-3 natural signs. At least some are type-1. But all type-3 natural signs that manifest the physical world are sensations.

Reid's theory of perception relies on conceptions and beliefs as components, perhaps marking the boundaries of the perceptual process. A successful perceptual process is initiated by properly caused sensations, that is, sensations arising from brain events linked to appropriately stimulated sense organs and ultimately to perceptual objects, which in turn trigger conceptions of and beliefs in the objects. The objects of conception just are the objects of perception, the referential contents. The ways in which conceptions characterise their objects are the presentational contents. When an unlearned perceiver experiences a sensation that prompts a perceptual conception that gives the perceiver an understanding of the thing's nature, the sensation is a type-3 natural sign. When it yields a lesser understanding, it is a type-1 natural sign. This is the basic philosophical scheme required to grasp Reid's primary/secondary quality distinction.

Thomas Reid's Primary/Secondary Quality Distinction

Philosophers traditionally contrast the secondary qualities of colour, sound, smell, taste and heat with the primary qualities of size, figure, solidity, divisibility and texture. What constitutes the distinction? Having mentioned clarification of the primary/secondary distinction as a top priority for his *Inquiry*, Thomas Reid addresses this question in at least three ways, and his answer stands in contrast to those offered by other philosophers associated with the topic, like Locke, Lucretius and Aristotle.[37] Consider Reid's take, starting with his view in terms of its comparison to other theories, then types of natural signs, and finally referential and presentational contents.

Today's readers have inherited two troublesome paragraphs from Reid that state his position explicitly while multiplying grief and frustration among Reid scholars. They appear in the *Essays on the Intellectual Powers of Man*. First,

> There appears to me to be a real foundation for the distinction; and it is this: That our senses give us a direct and distinct notion of the primary qualities and inform us what they are in themselves:

But of the secondary qualities, our senses give us only a relative and obscure notion. They inform us only that they are qualities that affect us in a certain manner, that is, produce in us a certain sensation; but as to what they are in themselves, our senses leave us in the dark.[38]

Second,

Thus I think it appears, that there is a real foundation for the distinction of primary from secondary qualities; and that they are distinguished by this, that of the primary we have by our senses a direct and distinct notion; but of the secondary only a relative notion, which must, because it is only relative, be obscure; they are conceived only as the unknown causes or occasions of certain sensations with which we are well acquainted.[39]

These strikingly repetitive paragraphs appear within a page of one another. Note the reappearance of 'real foundation', 'direct' and 'distinct', and 'relative' and 'obscure'. Despite the redundancy, it is not easy to grasp Reid's meaning. I confess that I misunderstood these passages several times before I understood them. What is Reid saying, and how does it relate to his broader corpus? One likely pitfall: it is easy to be misled by the appearance of the word 'relative' so near the term 'secondary qualities'. The careful reader should note that it is the notion that is relative and not the qualities. Secondary qualities are physical but also distinct from primary qualities.

In these passages, Reid seems to emerge as a unique voice in the philosophical community. Consider the logical space for a moment. Either secondary qualities are special, distinct from physical primary qualities, or not. Either secondary qualities are physical or not. Most early moderns, like the traditional primary/secondary guru John Locke, take secondary qualities to be mental and suppose them to be distinct from primary qualities in this respect. Idealists, like George Berkeley, regard both primary and secondary qualities as mental and thereby eliminate the distinction. Aristotle and Thomas Aquinas acknowledge only an insignificant distinction between primary and secondary qualities (which they call 'common sensibles' and 'special sensibles' respectively), but they take both types to be physical. This permits only one other view (Fig. 4.1) – Reid's.

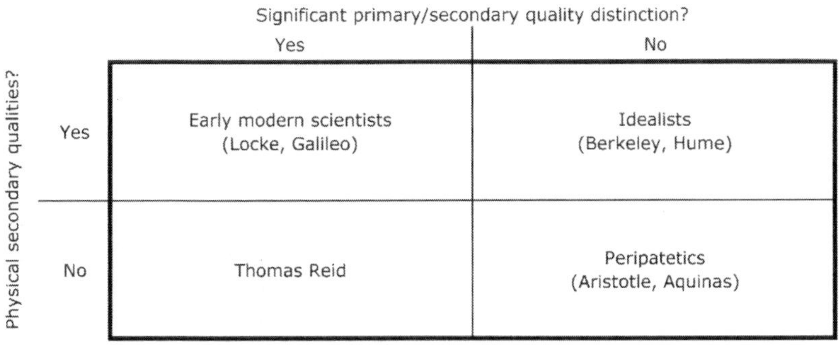

Significant primary/secondary quality distinction?

		Yes	No
Physical secondary qualities?	Yes	Early modern scientists (Locke, Galileo)	Idealists (Berkeley, Hume)
	No	Thomas Reid	Peripatetics (Aristotle, Aquinas)

Figure 4.1 Primary/secondary quality conceptual space

Opposite Idealism, Reid maintains the primary/secondary distinction as well as regards secondary qualities as physical properties. Again, he takes both types of quality as physical. But there is a distinction, based on the dissimilar ways in which human perceivers come to understand these qualities. Having distanced Reid's theory from other primary/secondary quality theories, you will no doubt ask what constitutes Reid's distinction between the quality types. Excellent question. Here is the answer in two additional ways.

First, in terms of natural signs. A nit-picker might characterise Reid's approach as a primary/secondary sensation distinction, rather than a quality distinction. Reid classifies qualities by the sorts of sensations they typically cause – type-1 versus type-3 natural signs. Recall from the last section that type-1 natural signs signify things as causes of the signs but are otherwise uninformative. Type-3 natural signs signify things as they are in themselves, according to their natures. For human perceivers, secondary qualities cause sensations that act as type-1 natural signs. As the above passage from Reid puts it, these sensations 'inform us only that they are qualities that affect us in a certain manner'. The natures of the sensations' causes remain hidden, at least initially. In other words, secondary qualities trigger sensations that prompt perceivers to recognise the qualities as causes of those sensations, not as anything else. Primary qualities also cause sensations, but these sensations act as type-3 natural signs. That is, they 'inform us what they are in themselves'. You automatically understand the nature of the primary quality upon experiencing the accompanying sensation. It is the type-3 signs, the primary quality sensations, that offer a transparent window to the external world and form the foundation of all human learning about the natures of physical things.

Buras's referential and presentational contents also help in explicating Reid's primary/secondary distinction. Recall that the referential content of a conception is its object, whereas the presentational content is the way that the conceiver considers the object. Conceptions may differ in referential contents while sharing presentational contents and vice versa. You might have six coffee mugs in your house and six corresponding conceptions that simply consider the mugs as mugs – different referential but similar presentational contents. On the other hand, try thinking of one particular mug as the recently emptied mug or as a gift from a relative. In this case, the two conceptions share a referential content, the mug, but their presentational contents vary – recently emptied vessel versus gift from family member. By now, it's clear that Reid regards conception as an essential component of perception. So, as conception-based mental operations, perceptions also have referential and presentational contents. How do primary and secondary qualities differ with regard to the sorts of contents involved in their perceptions?

On Reid's view, there is no significant difference in the sorts of referential contents involved, whether one perceives a primary or secondary quality. Both primary and secondary qualities are physical, so perceptions of both primary and secondary qualities take physical properties as objects. However, presentational contents offer a different story. When one naively perceives a primary quality, the presentational content characterises the quality with a 'direct and distinct notion'. If one were to put those conceptions into language, one might find oneself giving a scientific identity – conceiving of hardness as the firm adhesion of the parts of a body – or treating it as conceptually simple. Chemists may discover the cause of hardness, micro-physical chemical bonding, but they do not need to explain the nature of hardness. When one naïvely perceives a secondary quality, one finds its nature mysterious since one has only a 'relative and obscure notion'. The presentational contents of secondary quality perceptions bring them under relational conceptions, as the causes of such-and-such sensations – for example, sounds as causes of sound-type sensations. To acquire a more developed, scientific conception, you must either do some science yourself or consult a scientist.

When it comes to primary and secondary qualities, the senses yield different types of conceptions. Reid offers several examples that highlight the difference. Consider smelling a rose:

> Suppose a person who never had this sense [smell] before, to receive it all at once and to smell a rose . . . He finds himself affected in a new way, he knows not why or from what cause. Like a man that feels some pain or pleasure formerly unknown to him, he is conscious that he is not the cause of it himself; but cannot, from the nature of the thing, determine whether it is caused by body or spirit, by something near, or by something at a distance.[40]
>
> . . . a little experience will discover to him, that the nose is the organ of sense, and that the air, or something in the air, is a medium of it. And finding, by farther experience, that when a rose is near, he has a certain sensation; when it is removed, the sensation is gone; he finds a connection in nature betwixt the rose and this sensation. The rose is considered as a cause, occasion, or antecedent, of the sensation; the sensation as an effect or consequent of the presence of the rose.[41]

Reid cuts a sharp distinction between the mental smell sensation and the physical smell quality. You understand the purely phenomenological sensation clearly. Nowhere in the process of smelling the rose do you discover the nature of the smell quality, the property of the rose that produces the sensation.

The sensation the rose causes is a type-1 natural sign, signifying a certain quality in the rose. What quality does the sign signify? The quality in the rose that causes the sensation. Which quality is that? Long ago, the answer was wholly unknown and still is unknown to young children and certain of the uneducated. But now science has a ready answer. Reid reports,

> Natural philosophy informs us, that all animal and vegetable bodies, and probably all or most other bodies, while exposed to the air, are continually sending forth effluvia of vast subtilty . . . All the smell of plants, and of other bodies, is caused by these volatile parts, and is smelled wherever they are scattered in the air.[42]

A thorough understanding of the rose's smell comes by way of systematic inquiry, not sense perception alone. Before consulting the scientist, you have only an imperfect understanding of the smell, considering it by way of its effect, as the cause of your sensation, not as 'effluvia'.[43] The presentational content of the perceptual conception is relative to the sensing subject, presented as whatever caused that sensation. Eventually, you come to associate and even identify the cause as some quality of the rose, and this realisation allows you an additional conception of the smell, as whatever in the rose caused the sensation. The scientist may develop a more sophisticated conception, as effluvia emanating from the rose. All these conceptions, from the unknown cause of the sensation to the effluvia, have the same referential content – the tiny particles that constitute the rose's smell. They differ, however, in presentational content.

Reid makes a distinction between original conceptions, which come by perception alone, and acquired conceptions, which develop through experience, reason or science. Smell, taste, hearing and colour, Reid says, 'originally give us only certain sensations, and a conviction that these sensations are occasioned by some external object'.[44] The original conception of the rose's smell is indeed obscure and relative to a sensation. The sensation-relative conception hides the true nature of the smell, even though the conception takes the smell as its object. One may acquire additional conceptions of the smell, as whatever in the rose caused the sensation or as the effluvia of the rose. Even if these more sophisticated notions come to be deployed in perceptions of the rose's smell, the perceiving subject must first acquire them through reason and experience.

Compare the progression of conceptions of the rose's smell to the conception involved in a tactile perception of a primary quality, say, the hardness of a table. Reid claims that in the hardness case there is a sensation, just as with the rose. But the sensation originally triggers an absolute conception of hardness, as the firm adhesion of the parts of the table. Without science, without reason and extensive experience, one may come to know the nature of the quality of hardness. Perception alone is sufficient. Reid explains,

> There is, no doubt, a sensation by which we perceive a body to be hard or soft. This sensation of hardness may easily be had, by pressing one's hand against the table, and attending to the feeling that ensues, setting aside, as much as possible, all thought of the table and its qualities, of any external thing. But it is one thing to have the sensation, and another to attend to it, and make it a distinct object of reflection. We are so accustomed to use the sensation as a sign, and to pass immediately to the hardness signified . . .[45]

The sensation caused by the hardness of the table is a type-3 natural sign. What does the sign signify? As with the rose, the cause of the sensation. Which quality is that? The table's hardness – the firm adherence of its parts. It would be accurate to think of the table's hardness as whatever caused the sensation. However, in this case, such a relative and obscure conception is unnecessary. Type-3 sensations yield direct and distinct conceptions 'by a natural kind of magic', as a matter of common sense.[46] You understand the objective nature of the table's hardness immediately. You know what caused the sensation and do not need scientists to explain it. The presentational contents of an unlearned, perception-based conception and of the scientist's conception both present the hardness as the firm adhesion of the table's parts.

Make no mistake. The scientist is indispensable for helping to understand the cause of the hardness.[47] It is one thing to see that the table's parts adhere firmly, another to know that this is because the table is made of wood rather than metal or plastic. The scientist can testify about the wood, the cell-level features that make this type of wood particularly hard, or perhaps about the electrostatic and chemical forces that explain the cell-level features. But concerning the nature of the hardness, that it is the firm adherence of the table's parts. The scientist has no special knowledge or insight.

These examples illustrate the primary/secondary quality distinction in terms of natural signs and the contents of perception, but they also make clear what Reid means by 'direct and distinct' primary qualities and 'relative and obscure' secondary qualities. The notion of a primary quality that a typical human perceiver gains through perception features that quality's physical nature directly, without reference to anything else. Your notion of a secondary quality, if acquired in the normal sensory way, involves a reference to the sensation caused by that quality. It is a relative notion. As Reid puts it, 'A relative notion of a thing, is, strictly speaking, no notion of the thing at all, but only of some relation which it bears to something else'.[48] Moreover, a perception-based conception of a primary quality involves a higher level of understanding than that of a secondary quality, distinct (indeed, perfectly so) in comparison to the secondary quality notion's obscurity.

To reiterate an important similarity among primary and secondary qualities, notions of both are formed via the senses, with sensations acting as natural signs that trigger conceptions in perceivers. In every case, a physical quality affects sense organs to produce a sensation which leads the perceiver to conceive of and believe in the quality. Primary qualities are immediately understood by means of perception, both directly, in terms of their objective natures, and as physical properties of bodies. Because the perceiver understands primary qualities directly, she can offer scientific identifications of them. Most perceivers naturally attribute secondary qualities to physical objects as well, but it takes a great deal more effort to learn their natures. The naive perceiver conceives of them as unknown causes in some external object.[49] Thus, if Reid is correct, then physical objects possess secondary qualities, as common sense suggests. Understanding the natures of secondary qualities and giving scientific accounts of them, however, requires long-term observations and scientific effort.

Three concerns remain in Part II. First, scholars vary widely in their interpretations of Reid on primary and secondary qualities, and the reading proposed here is perhaps unique. So the remainder of Chapter 4 presents an extended defence of the present interpretation, showing plainly that this reading coheres not only with Reid's general statements about the primary/secondary distinction but also with Reid's discussions of primary and secondary quality species – extension, figure, hardness, solidity, colour, sound, smell and so on. After that, Chapter 5 gives an explicit response to the Problem of Secondary Qualities. Then, returning to the issue of understanding Reid's theory, Chapter 6 address several key interpretive questions – about the nature of the primary/secondary quality distinction, the perceptibility and causal efficacy of dispositions, and the application of the primary/secondary distinction to properties other than those explicitly discussed by Reid.

The Primary Qualities

Unfortunately, Reid nowhere undertakes an encyclopaedic account of the primary and secondary qualities, explaining how his primary/secondary distinction applies to each quality type. Instead, Reid's reader finds comments like,

> What hath been said of hardness, is so easily applicable, not only to its opposite, softness, but likewise to roughness and smoothness, to figure and motion, that we may be excused from making the application, which would only be a repetition of what hath been said.[50]

And

> The same reasoning [as was just applied to smell] will apply to
> every secondary quality.[51]

But perhaps Reid is wrong to think that a careful reader can
automatically fill in the implications of his primary/secondary
distinction. If it were as easy as Reid suggests, then someone
would have done it. No one has. Perhaps Reid's interpreters also
regard it as an exercise best left to individual readers. If so, then
this too is a mistake. This exercise left to the reader may be partially to blame for the wide variety of misreadings of Reid on
primary and secondary qualities discussed in Chapter 6. Here,
the remainder of Chapter 4 does what no one else has bothered
to do. It accounts for Reid's take on particular species of primary
and secondary qualities and various human conceptions of them
in a case-by-case fashion. Ultimately, this exercise settles several
key interpretive questions about Reid's primary/secondary distinction and its implications.

Start with the primary qualities, which Reid names as extension, divisibility, figure, motion, solidity, hardness, softness, and
fluidity.[52] The following analysis identifies the various presentational contents that Reid attributes to each conception of the primary qualities. Where possible, it identifies the sensations involved
in original perceptions of them.

Hardness, Softness and Fluidity

In the previous section, hardness served as the paradigm case for
a primary quality perception, so it has already been covered. The
words 'hardness' and 'softness' indicate specific types of qualities
possessed by physical objects. Reid claims that when you perceive
the hardness of an object via your sense of touch, you form a clear
and distinct conception of it.[53] That is, there is no obscurity in the
presentational content of the conception formed, because the clear
conception involves no reference to sensations or other mediating
objects or properties. The conception is also direct, comprehend-
ing the object as it is in itself.[54] Reid explicitly identifies the presentational content of that conception as 'the cohesion of the parts
of a body with more or less force'.[55] He later attaches this presentational content to fluidity as well.[56] In the next line, Reid points

out that this conception does not reveal the causal reason for those parts cohering firmly. Figuring out causes is the task of science. Science may also discover precise ways to measure and compare the hardness of various objects. But there is no call for a scientific inquiry into the nature of the hardness, since its nature as the firm adherence of an object's parts is manifest in sense perception.

The presentational content of the perceptual conception of hardness as cohesion with more or less force yields a definition of hardness, but not just any definition. It is the sort that Reid calls 'logical' and others call an analytic specification, following the language of Hilary Putnam.[57] This definition explains what it is to be hard in terms of its conceptual parts. Again, the primary quality of hardness, when perceived via touch, involves a sensation. Reid claims that there is no name for that sensation, but he can form a clear conception of it.[58] He says,

> The notion of hardness in bodies, as well as the belief of it, are got in a similar manner; being, by an original principle of our nature, annexed to that sensation which we have when we feel a hard body. And so naturally and necessarily does the sensation convey the notion and belief of hardness, that hitherto they have been confounded by the most acute inquirers into the principles of human nature, although they appear, upon accurate reflection, not only to be different things, but as unlike as pain is to the point of a sword.[59]

There is a sensation involved in perceiving hardness by touch, different from both the conception of hardness and the quality of hardness. Plus, the perception-based conception of hardness does not present that quality in terms of any relation it bears to the sensation it causes. It is a direct conception; the sensation makes no appearance in the presentational content of the conception. If it did, you would be 'in the dark' about the quality's perceiver-independent nature.[60]

Solidity

Reid gives the presentational content of his notion of solidity: 'It excludes other bodies from occupying the same place at the same time'.[61] Strikingly, Reid identifies solidity as a disposition, that is, by what objects with that quality tend to do. Like the conception of hardness, it is an analysis – direct and distinct. You know exactly what it is because you understand the disposition.[62] Reid doesn't discuss the sensation associated with solidity perceptions, but his general description of the primary qualities indicates that there is a sensation. As you can see from Reid's description, the presentational contents in perception-based notions of solidity do not make reference to the sensations caused by the quality.[63]

Extension, Figure, Motion and Divisibility

Tactile perceptions of the primary quality of hardness involve a certain sensation. The same sensation, Reid says, also triggers conceptions of three other bodily qualities – extension, real (as opposed to visible) figure, and relative (as opposed to absolute) motion:[64]

> Extension, therefore, seems to be a quality suggested to us, by the very same sensations which suggest the other qualities above mentioned. When I grasp a ball in my hand, I perceive it at once hard, figured, and extended. The feeling is very simple, and hath not the least resemblance to any quality of body. Yet it suggests to us three primary qualities perfectly distinct from one another, as well as from the sensation which indicates them.[65]

Reid adds motion to this list a few lines later. Thus, there is a single tactile sensation that Reid associates with extension, real figure, relative motion and hardness. With practice, most perceivers can learn to recognise extension, real figure and relative motion visually and even aurally.[66] But original perceptions apprehend them via touch.

Here would be the place to submit Reid's analyses of these qualities for your examination, but he makes no great effort to translate the presentational contents of conceptions of extension, figure or motion into language. By real figure Reid means a physical object's shape, a quality perfectly understood even to the blind mathematician Dr Saunderson, but 'shape' is a synonym for 'real figure', not an analysis.[67] Reid thinks that all motion, whether relative or absolute, presupposes some fixed point.[68] But, again, this falls short of saying what motion is. Finally, Reid thinks that extension, figure, hardness or softness, roughness or smoothness, and motion or rest link up conceptually such that a body cannot have any one without the others.[69] Still, there is no scientific identity here, no explanation of what they are in themselves.

It may seem surprising that Reid does not offer scientific identities for these qualities, but perhaps it's because he takes them to be conceptual simples. Where one would expect Reid to define belief, he refuses, saying, 'The operation of the mind signified by [the words that indicate belief] is perfectly simple, and of its own kind.'[70] Reid regards belief as not subject to conceptual analysis. Likewise, he thinks that he, Dr Saunderson, and other adults have clear and distinct conceptions of extension, figure, motion and perhaps divisibility even without logical definitions.[71] Most likely, divisibility is a disposition to divide in certain circumstances. But, otherwise, Reid finds no need for logical or strict and proper definitions, since everyone understands exactly what these qualities are. In fact, they are the properties by which other properties are explained. As Reid puts it, 'The distinctness of our notions of them enables us to reason demonstratively about them to a great extent.'[72] Although one may note certain conceptual relationships of each quality to the others, there is no reason to break these properties into their conceptual parts. Each is perfectly understood as is.[73]

Reid leaves untreated the sensations involved in perception of divisibility.[74]

The Secondary Qualities

Reid thinks that perceivers readily understand each of the primary qualities, so there is no hypothesising about their natures. However, Reid finds secondary qualities appropriate subjects for speculation.[75] In these cases, the scientist may inquire not only into the causes and effects of the qualities but also their natures. Reid encourages his reader to offer and defend various hypotheses about what the secondary qualities are, and in fact scientists often entertain more than one proposal when debating their natures. The following sections, then, note not only what Reid thinks the secondary qualities are but what he thinks they might be. That is, these conceptions consider the presentational content of each conception as given by sense perception alone as well as by speculative conceptions, including inaccurate ones, with the same referential content.

Smell

Returning to the smell of the rose, the presentational content of a purely perceptual conception is not very rich. Reid explores it in a thought experiment:

> Suppose a person who never had this sense before, to receive it all at once, and to smell a rose; can he perceive any similitude or agreement between the smell and the rose? Or indeed between it and any other object whatsoever? Certainly he cannot. He finds himself affected in a new way, he knows not why or from what cause. Like a man that feels some pain or pleasure formerly unknown to him, he is conscious that he is not the cause of it himself; but cannot from the nature of the thing, determine whether it is caused by body or spirit. It has no similitude to any thing else, so as to admit of a comparison; and therefore he can conclude nothing from it, unless perhaps that there must be some unknown cause of it.[76]

The original conception of the rose's smell arises upon experiencing a certain sensation, but this type-1 sensation lacks the conceptual magic of the primary qualities' type-3 sensations. So the sensation inclines you to think of the smell as just the quality that caused the sensation. You also understand the smell quality as mind-independent, since you did not cause the sensation.[77]

Reid extends his consideration of how you might conceptualise the smell, outlining the roles that experience, memory and induction might play in acquiring more sophisticated conceptions. He continues,

> A little experience will discover to him, that the nose is the organ of this sense, and that the air, or something in the air, is a medium of it. And finding, by farther experience, that when a rose is near, he has a certain sensation; when it is removed, the sensation is gone; he finds a connection in nature betwixt the rose and the sensation. The rose is considered as a cause, occasion, or antecedent, of the sensation; the sensation as an effect or consequence of the presence of the rose.[78]

Experience allows you an additional conception of the rose's smell. The presentational content of this new conception is 'the quality in this rose that causes me to experience that sensation'. This makes two conceptions of the smell – one from sense perception and the other from sense plus experimentation with the rose. The two differ only with regard to their presentational content, not their referential content, and even then only slightly.

A third conception becomes available through scientific investigation, when you learn that smelly things smell, in part, because they emit tiny particles or 'effluvia' into the air that get drawn in by the perceiver's nose.[79] This discovery reveals a conception that possesses yet another presentational content, 'some power, quality, or virtue, in the rose, or in effluvia proceeding from it, which hath a permanent existence, independent of the mind, and which, by the constitution of nature, produces the sensation in us'.[80] Although not a scientific identity, this conception constitutes a clear advance in understanding over the initial conception.[81]

Nothing about the smell sensation alone or the presentational content of the initial conception determines a right-thinking person to adopt this third or even the second conception of the rose's smell. According to Reid, scientific investigation could have led, epistemically speaking, to the view that the smell resulted from something spiritual rather than physical.[82] He mentions a theory of smell on which each body has a soul that 'flies about in the air in quest of a proper receptacle'.[83] He considers vibrations as a plausible theory of smell.[84] For all anyone knows by sense perception alone, the cause of the sensation might be one of these things. So the third conception, involving effluvia, is theoretical and a posteriori, a well-evidenced hypothesis that aims to comprehend the true nature of the smell and perhaps succeeds.

Taste

Reid does not consider a detailed case of tasting, as he does with smelling. But the fact that he identifies it as a secondary quality shows his opinion on the presentational content of your initial conception of some taste: 'the cause of that taste sensation'. With experience, you soon discover that the sensation happens when placing an object in your mouth or on your tongue. So you can acquire a second, more experienced conception of the taste with a presentational content that takes this into account, akin to the second conception of the smell in the last section.

Reid includes a few notes that might lend themselves to the development of more scientific conceptions. He says, for example, that the quality has something to do with an object's being soluble in saliva.[85] He also notes a correlation between taste and smell, as good smelling things are likely to be good tasting. From these discoveries, he hopes, scientists will develop other more analytical conceptions, a way to understand taste perfectly.

Sound

By means of sense perception, the secondary quality of sound grants a conception with the presentational content, 'the cause of that aural sensation'. Of course, you learn to associate variations sensations with types and relative locations of sounding objects. In other words, as in the cases of smell and taste, you naturally acquire additional conceptions of sounds by means of everyday experiences – for example, 'the ringing sound of that bell' or 'the barking sound of her dog'. Also like the previous cases, you can develop more sophisticated conceptions of sound by way of everyday experience and scientific discovery. Reid says that scientists have discovered a medium involved in perceptions of sounds. He explains,

> To make a perfect tone, a great many undulations of elastic air are required, which must all be of equal duration and extent, and follow one another with perfect regularity; and each undulation must be made up of the advance and recoil of innumerable particles of elastic air, whose motions are all uniform in direction, force, and time.[86]

Just as effluvia serve as a medium for smells and saliva for tastes, vibrating air enables human hearers to hear sounds.[87]

Unlike smell and taste, Reid thinks that scientific inquiry has succeeded in developing a direct and distinct conception, and thereby an accurate scientific identification, for sound. The vibrations of the air communicate the relevant quality of the sounding body. The body's vibration causes the vibrations in the air. Thus, a thing's sound is its vibration.[88] This identification appears more than once in Reid's writing. For example, he compares the vibration of a bell to the sensation it causes precisely to show that perceived qualities, like bell vibrations, do not resemble sensations.[89] He also draws an analogy between the relation of hardness to the conceptual presentational content, 'firm adhesion of the parts of a body', and sound to the presentational content, 'vibration in the sounding body'.[90] Reid takes the presentational content of this scientific conception of sound as direct and distinct in exactly the way the perceptual conception of hardness is. The two conceptions equally reveal the objective natures of their objects. Sound can be understood as directly and distinctly as hardness. However, acquiring a conception of sound as a vibration requires significant intellectual work, whereas the full understanding of hardness comes automatically via sense perception.

Heat and Cold

According to Reid, your original, perception-based conception of heat resembles those of smell, taste and sound. It has a presentational content of the form, 'the cause of that sensation', and becomes part of your thought life when you touch a hot or cold object. Further conceptions begin to flesh out a more detailed account of heat as a quality of physical objects. Reid reiterates that common sense insists on treating it as objective, even though its exact physical nature is unknown, saying, 'For what could be more absurd, than to say, that the thermometer cannot rise or fall, unless some person be present, or that the coast of Guinea would be as cold as Nova Zembla, if it had no inhabitants?'[91] That is, the conception of heat developed through everyday experience incorporates objectivity as part of its presentational content.

Reid is in the dark about the objective natures of smells and tastes, fully informed about sounds, and between lost and found on hot and cold. He offers no definite analysis of heat but instead ventures a few guesses. Reid writes his *Inquiry* in 1763, when the phlogiston theory of heat, although fatally wounded by Mikhail Lomonosov, still dominates the theoretical-scientific landscape. Phlogiston's death blow comes with Antoine Lavoisier's 'Reflections on Phlogiston' in 1783. No doubt, Reid means to contrast the phlogiston theory with his own suggestion, 'a certain vibration of the parts of the heated body'. Furthermore, Reid raises questions about heat's relation to cold – whether the two are contraries or one the privation of the other.[92] He expects one of these options to become incorporated into the scientist's conception of heat. But, for Reid, actually adopting one view or the other would be too hasty. Logical definitions of these qualities, and thus direct and distinct conceptions, require further investigation.

Colour

As with other secondary qualities, Reid keenly distinguishes colours from the sensations they cause, which he calls appearances. Again, Reid holds that the connection between colour appearance and colour is merely contingent: 'Although there is no resemblance, nor, as far as we know, any necessary connection, between that quality in a body which we call its colour, and the appearance which that colour makes to the eye . . .'[93] But sight originally offers only one conception of colour, as the cause of the appearance and perhaps as possessed by a body.[94]

As with smell, taste, sound and heat, Reid's theory suggests that a normal perceiver develops several colour conceptions, all of which have the same referential content, a colour, but differing presentational contents. According to Reid, some conceptions gained by experience involve presentational contents that take proper viewing conditions into account, 'a certain power or virtue in bodies, that in fair day-light exhibits to the eye an appearance, which is very familiar to us, although it hath no name'.[95] Note that this content includes the usual placement of the quality in a physical body. It also accompanies the realisation that light serves as the medium by which colour makes its causal impression on the eye.[96]

Reid includes a modest proposal for an analysis of colour: 'the disposition of bodies to reflect a particular kind of light'.[97] But his analysis goes no further. Reid offers no species into which one might sort these particular kinds of light. However, it is telling that this is the only secondary quality that Reid identifies as a disposition. So Reid has the beginnings of an identity conception of colour as a disposition, perhaps a spectral reflectance. The peculiar place of colour for Reid's theory receives further attention in Part III.

All that has been observed here should make Reid's view more perspicuous. His distinction between primary and secondary qualities is single faceted and anthropological. Jennifer McKitrick characterises it as 'epistemic'.[98] Human conceptions of primary qualities formed solely on the basis of unlearned sense perceptions are direct and distinct. By sense perception alone you can understand the what-it-is of primary qualities. Sometimes these direct and distinct conceptions allow you to scientifically identify the quality, as in the cases of hardness and solidity. Other times, perception-based understandings are conceptually simple but nonetheless fully understood, as in extension and figure. Original sense perceptions of secondary qualities, on the other hand, involve indirect and obscure conceptions. In particular, original perceptions present secondary qualities as causes of the sensations that they happen to produce in you. This is because the sensations by which you learn about secondary qualities are type-1 natural signs, which means that you must learn the connection between the sign and the thing signified through other means. Original perceptions do not reveal the what-it-is of the quality. They merely say how the quality relates to certain sensations, as their causes. Such conceptions are relative and obscure because their presentational contents reference the sensation rather than addressing the quality's objective nature.

The limited knowledge from secondary quality sense perceptions alone highlights Reid's love of science and everyday experience, through which perceivers develop additional, more informative conceptions. He tasks scientists with discovering scientific identities for secondary qualities. Thus, his primary/secondary distinction is committed to secondary qualities as physical, objective properties of bodies, just as common sense suggests. Primary and secondary qualities are, metaphysically speaking, equally real. Reid writes in a letter to Lord Kames (Henry Home),

> What is in the Sugar then. Not the Sensation but what causes the Sensation in us. The word Taste is applied both to the Sensation and to the Quality in the body tasted, & most frequently by the Vulgar to the last. Therefore if we will use common Words in their common Acceptation, which I think Philosophers ought to do, we both speak properly and think justly when we say that Color Sound and Taste are in the external Objects.[99]

According to Reid, secondary qualities are just as physical and objective as primary qualities. The true demarcation between primary and secondary concerns only the way in which human perceivers understand them, not their essential natures.

Notes

1. Reid, *Essays on the Intellectual Powers of Man*, 2.5/96.
2. Van Cleve, *Problems from Reid*, p. 14; Reid, *Inquiry into the Human Mind*, 6.21/174; Reid, *Essays on the Intellectual Powers of Man*, 2.2/76.
3. Reid, *Essays on the Intellectual Powers of Man*, 1.1/24–5.

4. Reid, *Essays on the Intellectual Powers of Man*, 6.1/406–7.
5. Reid, *Essays on the Intellectual Powers of Man*, 2.20/228.
6. Reid, *Essays on the Intellectual Powers of Man*, 2.20/227–8.
7. See Reid, *Essays on the Intellectual Powers of Man*, 2.5/98.
8. Reid, *Essays on the Intellectual Powers of Man*, 2.5/97.
9. Pelser, 'Belief in Reid's Theory of Perception'.
10. Reid, *Essays on the Intellectual Powers of Man*, 2.8/126; Copenhaver, 'Thomas Reid's Direct Realism'.
11. Reid, *Essays on the Intellectual Powers of Man*, 2.18/214.
12. Reid, *Essays on the Intellectual Powers of Man*, 4.2.17–20/321.
13. Reid, *Essays on the Intellectual Powers of Man*, 1.1/22–3.
14. Buras, 'The Problem with Thomas Reid's direct realism', pp. 50–1; Buras, 'The Function of Sensations in Reid'.
15. Smith, *The Problem of Perception*, pp. 94–121.
16. Hanson, *Patterns of Discovery*, p. 4.
17. Dretske, 'Perception without Awareness'.
18. Van Cleve, *Problems from Reid*, pp. 13–21.
19. Reid, *Essays on the Intellectual Powers of Man*, 1.1/25.
20. Reid, *Inquiry into the Human Mind*, 2.2/26; *Essays on the Intellectual Powers of Man*, 2.17/201.
21. Van Cleve, *Problems from Reid*, p. 15.
22. Reid, *Essays on the Intellectual Powers of Man*, 1.1/37.
23. This is not to say that sensations are without purpose for Reid. They serve as triggering mechanisms for perceptions. And they are also useful in making one feel (Reid, *Essays on the Intellectual Powers of Man*, 2.18/210). This second feature, Reid thinks, is a survival-conducive one, so there is a biological and perhaps even evolutionary advantage to experiencing sensations. See also Nichols, *Thomas Reid's Theory of Perception*, pp. 150–2.
24. Some have thought that Reid believes living, embodied humans do sometimes perceive visible figure without sensations. See Tebaldi, 'Thomas Reid's Refutation of the Argument from Illusion' and Nichols, *Thomas Reid's Theory of Perception*, p. 169. As explained later in this chapter, these are misinterpretations.
25. Reid, *Essays on the Intellectual Powers of Man*, 1.1/23.
26. Hume, *Treatise on Human Nature*, 1.1.1.1/7.
27. Reid, *Essays on the Intellectual Powers of Man*, 1.1/23.
28. Reid, *Inquiry into the Human Mind*, 6.20/167–8.
29. Buras, 'The Nature of Sensations in Reid'.
30. Reid, *Inquiry into the Human Mind*, 6.24/190.
31. Reid, *Inquiry into the Human Mind*, 5.3/58.
32. Descartes, *Optics*, 165.
33. Reid, *Inquiry into the Human Mind*, 5.3/59.
34. Reid, *Inquiry into the Human Mind*, 5.3/58–61.

35. Reid, *Inquiry into the Human Mind*, 5.3/60.
36. Reid, *Inquiry into the Human Mind*, 5.4/61; Reid, *Inquiry into the Human Mind*, 5.7/70.
37. Reid, *Inquiry into the Human Mind*, Abstract/259.
38. Reid, *Essays on the Intellectual Powers of Man*, 2.17/201.
39. Reid, *Essays on the Intellectual Powers of Man*, 2.17/202.
40. Reid, *An Inquiry into the Human Mind*, 2.2/26.
41. Reid, *Inquiry into the Human Mind*, 2.8/39–40.
42. Reid, *Inquiry into the Human Mind*, 2.1/25.
43. Reid, *Essays on the Intellectual Powers of Man*, 2.17/202.
44. Reid, *Essays on the Intellectual Powers of Man*, 2.21/234–41.
45. Reid, *Inquiry into the Human Mind*, 5.2/55–6; Reid, *Essays on the Intellectual Powers of Man*, 2.17/204.
46. Reid, *Inquiry into the Human Mind*, 5.4/60–1.
47. Reid, *Inquiry into the Human Mind*, 5.5/61.
48. Reid, *Essays on the Intellectual Powers of Man*, 2.17/201.
49. Reid, *Essays on the Intellectual Powers of Man*, 2.21/235.
50. Reid, *Inquiry into the Human Mind*, 5.4/62.
51. Reid, *Essays on the Intellectual Powers of Man*, 2.17/202.
52. Reid, *Essays on the Intellectual Powers of Man*, 2.17/201; Reid, *Inquiry into the Human Mind*, 5.1/54.
53. Reid, *Inquiry into the Human Mind*, 5.4/61, 5.6/63–5.
54. Reid, *Essays on the Active Powers of Man*, 1.1/9; Reid, *Essays on the Intellectual Powers of Man*, 2.17/201.
55. Reid, *Inquiry into the Human Mind*, 5.4/61.
56. Reid, *Essays on the Intellectual Powers of Man*, 2.17/201.
57. Reid, *Essays on the Intellectual Powers of Man*, 2.20/227.
58. Reid, *Inquiry into the Human Mind*, 5.4/60–2.
59. Reid, *Inquiry into the Human Mind*, 5.3/60; Reid, *Essays on the Intellectual Powers of Man*, 2.17/204–5.
60. Reid, *Essays on the Intellectual Powers of Man*, 2.17/201.
61. Reid, *Essays on the Intellectual Powers of Man*, 2.17/201.
62. Prior et al., 'Three Theses about Dispositions'.
63. See also Reid, *Essays on the Intellectual Powers of Man*, 2.17/202.
64. Reid, *Essays on the Intellectual Powers of Man*, 2.22/245.
65. Reid, *Inquiry into the Human Mind*, 5.5/63.
66. Reid, *Inquiry into the Human Mind*, 6.2/79; Reid, *Essays on the Intellectual Powers of Man*, 2.21/236.
67. Reid, *Inquiry into the Human Mind*, 5.6/65.
68. Reid, *Essays on the Intellectual Powers of Man*, 2.22/246.
69. Reid, *Inquiry into the Human Mind*, 5.5/62.
70. Reid, *Essays on the Intellectual Powers of Man*, 2.20/227.
71. Divisibility appears on a list with extension, figure and motion in Reid, *Essays on the Intellectual Powers of Man*, 2.17/201. Saunderson is mentioned in Reid, *Inquiry into the Human Mind*, 5.6/65.

72. Reid, *Essays on the Intellectual Powers of Man*, 2.17/203.
73. Reid, *Essays on the Intellectual Powers of Man*, 2.17/201.
74. Reid, *Essays on the Intellectual Powers of Man*, 2.19/219.
75. Reid, *Essays on the Intellectual Powers of Man*, 2.17/204.
76. Reid, *Inquiry into the Human Mind*, 2.2/26; Reid, *Essays on the Intellectual Powers of Man*, 2.21/235.
77. Reid, *Inquiry into the Human Mind*, 2.2/26.
78. Reid, *Inquiry into the Human Mind*, 2.8/39–40.
79. Reid, *Inquiry into the Human Mind*, 2.1/25.
80. Reid, *Inquiry into the Human Mind*, 2.9/43.
81. Reid seems inconsistent on whether he regards effluvia as the smell itself or merely the smell's medium. Compare Reid, *Inquiry into the Human Mind*, 5.1/25, 5.4/61 and Reid, *Essays on the Intellectual Powers of Man*, 2.17/204 to Reid, *Inquiry into the Human Mind*, 6.21/174.
82. Reid, *Inquiry into the Human Mind*, 2.2/26.
83. Reid, *Inquiry into the Human Mind*, 2.1/25.
84. Reid, *Inquiry into the Human Mind*, 5.2/57.
85. Reid, *Inquiry into the Human Mind*, 3/46.
86. Reid, *Inquiry into the Human Mind*, 4.1/49.
87. Reid, *Inquiry into the Human Mind*, 6.21/174.
88. Reid, *Essays on the Intellectual Powers of Man*, 2.17/204, 209.
89. Reid, *Essays on the Intellectual Powers of Man*, 2.17/203.
90. Reid, *Essays on the Intellectual Powers of Man*, 2.17/209.
91. Reid, *Inquiry into the Human Mind*, 5.1/54.
92. Reid, *Inquiry into the Human Mind*, 5.1/55.
93. Reid, *Inquiry into the Human Mind*, 6.7/95.
94. Reid, *Essays on the Intellectual Powers of Man*, 2.2.21/236.
95. Reid, *Inquiry into the Human Mind*, 6.5/57.
96. Reid, *Inquiry into the Human Mind*, 6.21/174.
97. Reid, *Essays on the Intellectual Powers of Man*, 2.17/204.
98. McKitrick, 'Reid's Foundation for the Primary/Secondary Quality Distinction', p. 73.
99. Reid, *The Correspondence of Thomas Reid*, 20 December 1778, 61/115.

5

Answering the Problem of Secondary Qualities

In Part I, the most vulnerable premise of the Problem of Secondary Qualities seemed to be the Secondary Quality Non-Physicality Thesis (NPT), that if an object is physical, then it does not possess secondary qualities. Robinson characterised this claim as a deliverance of science, citing Galileo, Descartes and Locke as authorities, and commitments to NPT appeared in the writings of contemporary scientists. In spite of philosophy's high regard for science and strong desire to accommodate its best theories, acceptance of NPT needed a more detailed argument. Frank Jackson offered such an argument, one that rests on a hypothesis regarding the causal potency of secondary qualities. Scientific properties, Jackson said, fully account for the causal connections between physical properties and sense organs. So sensations offer no reason to posit non-scientific properties in physical objects, anything in addition to scientific properties would be causally superfluous. This creates a problem for secondary qualities, said Jackson, because secondary qualities, being conspicuously absent from current scientific theories, are not scientific. Sensations do not count as evidence for the possession of secondary qualities by physical objects. Without the evidence of sense, why think that physical objects possess secondary qualities at all?

Rejecting the Non-Physicality Thesis

Rejecting NPT means challenging Jackson's claim that secondary qualities are not scientific and therefore not causally potent. How can secondary qualities be causally potent and therefore scientific? They seem not to be because they are not readily understood in scientific terms. Jackson is right that the language of science does not mention secondary qualities by their traditional names, and the conceptions by which perceivers originally grasp secondary qualities are not scientific notions. But these facts about secondary quality names and notions do not demonstrate anything about the natures of secondary qualities themselves, only the way that English speakers tend to think and talk about them. Even if traditional secondary quality vocabulary words do not appear in scientific explanations, it may be that scientific theories make reference to secondary qualities. This would be the case, for example, if secondary qualities were identical to known scientific properties or combinations of scientific properties.

Just as Robinson regards NPT as a discovery of science, Reid takes the falsity of NPT as scientific and commonsensical. That is, both Reid and Robinson regard the question of whether physical objects possess secondary qualities as an empirical one. Both believe that the question has been answered. But they arrive at different conclusions. On Reid's account, secondary qualities are identical with objective, causally potent properties of physical objects.

According to Reid, perceivers initially understand secondary qualities by way of their accidental features, like tendencies to cause certain sensations. Original, perception-based conceptions of secondary qualities have obscure and relative presentational contents, like 'the cause of that sensation'. You may think that the sensation was caused by some external object, since you know that you did not purposely cause it.[1] But, by your original conception, you cannot identify the external object as either body or spirit.[2] Initially, you only know the secondary quality as the cause of a sensation, not as a property of a physical object.[3] Again, this matter is empirical for Reid, just as it is for Robinson.

As outlined in Part I, the Problem of Secondary Qualities, the conclusion that secondary qualities are not physical properties, depends on two claims from Jackson: first, that to perceive properties of physical objects, they must be causally potent and scientific; second, that secondary qualities are not scientific, because scientific theories do not mention them. Reid reaches the opposite conclusion almost as quickly:

> [T]he nose is the organ of this sense, and that the air, or something in the air, is a medium of it. And finding, by farther experience, that when a rose is near, he has a certain sensation; when it is

removed, then sensation is gone; he finds a connection in nature betwixt the rose and this sensation. The rose is considered a cause, occasion, or antecedent, of the sensation.[4]

For Reid, it is a matter of common sense that secondary qualities are possessed by physical objects, a truth discovered by recognising that sensations are constantly conjoined to physical bodies, like roses and noses. You may not understand the natures of secondary qualities, but, according to Reid, you quickly learn that they are causally potent properties of certain physical objects.

Original conceptions of secondary qualities communicate only that secondary qualities are causally potent. Experience then locates secondary qualities in physical objects. Discovering anything else about them requires more extensive study. According to Jackson, further study of the causal connections between sense organs and the external world has no need for properties beyond those of current science.[5] If Jackson is correct here, then secondary qualities must be identical to today's scientific properties or combinations of today's scientific properties.

Jackson claims that secondary qualities are not scientific because their traditional names, like 'red', 'yellow', 'sweet' or 'concert A', do not appear in contemporary scientific descriptions of the world. But Reid would find Jackson's dismissal entirely too hasty. If perceivers know anything about secondary qualities, it is that they are scientific, since only scientific properties are causal. Given only that secondary qualities are causal properties of physical objects, it remains a matter for empirical investigation to say what kinds of secondary qualities there are and how they may be identified with scientific properties.

On Reid's account, secondary quality names attach to secondary qualities via original, relative and obscure conceptions. Perceivers may tell blue from scarlet, he says, only by the differences in the sensations that they cause.[6] Relative and obscure conceptions don't suit scientific reasoning. Reid remarks,

The primary qualities are the object of the mathematical sciences; and the distinctness of our notions of them enables us to reason demonstratively about them to a great extent. Their various modifications are precisely defined in the imagination, and thereby capable of being compared, and their relations determined with precision and certainty.[7]

Science prefers clear and distinct conceptions and vocabulary. Original conceptions of secondary qualities fall short of this requirement, even if secondary qualities are scientific properties. To think and speak of secondary qualities in a scientific way requires clear and distinct conceptions of those qualities.

Whose job is it to provide clear and distinct conceptions of secondary qualities? Reid charges scientists with the task. Perception needs help, so Reid trusts science to pick up where perception leaves off. He calls for scientists to investigate the natures of secondary qualities.[8] As seen earlier, he also believes that scientists have made significant progress to this end already, especially regarding sound and perhaps heat by the end of the eighteenth century. Initially, all anyone knows of secondary qualities are their sensational causal effects, mere accidental features. To reason clearly about secondary qualities, you need to know what they are. Reid asks scientists to study secondary qualities in order to develop more sophisticated, probing, scientific conceptions of them. Ultimately, scientists should make a posteriori identifications of secondary qualities with their natures. This is science's contribution to the grander philosophical project.

If Reid is right, then Jackson is wrong to say that secondary qualities are not scientific properties. They are scientific, but not yet fully understood in scientific terms (except for sound, which is a vibration of a body). If this is so, then Jackson is also wrong about the causal potency of secondary qualities. In fact, secondary qualities do affect sense organs, cause sensations, and thereby generate evidence for their physical existence. If secondary qualities are physical-causal, scientific properties, then they are possessed by physical objects and NPT is false. Without this premise, the Problem of Secondary Qualities cannot conclude that Direct Realism is false or that common sense has failed.

Reid and Kripke on A Posteriori Identity

Reid effectively says that secondary qualities are scientific, causal properties of physical bodies, even if no one knows precisely which secondary quality is identical to which scientific property. He charges scientists with making such identifications on the basis of empirical investigation. A posteriori identification, of the sort demanded by Reid, is a familiar exercise in contemporary analytic philosophy, thanks to writers like Saul Kripke and Hilary Putnam.[9] Don't draw a historical connection where there is none, but noting a few similarities between Reid and these thinkers may help considering Reid's proposal in light of recent scholarship.

Reid says that perceivers identify secondary qualities by the sensations that they cause. For example, Reid says of heat,

> The sensations of heat and cold are perfectly known; for they neither are, nor can be, any thing else than what we feel them to be; but the qualities in bodies which we call heat and cold, are unknown. They are only conceived by us, as unknown causes or occasions of the sensations to which we give the same names.[10]

You experience a sensation and conceive of and believe in some quality that is the cause of that sensation. Your conception of that quality has the sensation's cause as its referential content and 'the cause of this sensation' as its presentational content. English speakers call the referential content heat.

Reid sees the causing of particular sensations as merely accidental. Heat can exist without causing sensations.[11] Moreover, each secondary quality might have caused sensations other than the ones that it causes: 'No man can give a reason, why the vibration of a body [i.e., heat] might not have given the sensation of smelling, and the effluvia of bodies affected our hearing'.[12] Heat sensations merely pick out the heat quality for naive perceivers. They don't communicate the nature of heat. Furthermore, the name, 'heat', picks out the quality that causes heat sensations. It isn't an abbreviation for 'that which causes heat sensations'.

According to Kripke, humankind was for a long time in a similar situation with regard to gold. Experts identified gold by certain identifying marks, like its yellow colour.[13] This metal received the name 'gold'. But yellowness is not an essential property of gold. It is conceivable, says Kripke, that science discovers gold to be not yellow but blue or perhaps that there are other non-gold metals with most or all of gold's identifying marks – fool's gold.[14] Yellowness and other identifying marks help pick out gold, but they do not reveal the nature of gold. The name 'gold', being a rigid designator, is not an abbreviation for 'yellow metal', nor is being yellow metal part of what it means to be gold.[15]

According to Kripke, to understand the nature or essential properties of gold, you need to investigate it beyond its yellowness.

What properties of gold make it gold? Kripke suggests having the atomic number of 79. It seems that anything with atomic number 79 is gold and that gold is anything with atomic number 79.[16] It is part of the nature of gold to have atomic number 79. And understanding gold as the element with atomic number 79 represents a successful a posteriori identification of gold with its nature.

Conveniently, heat is another of Kripke's paradigm cases. For Kripke, the word 'heat' picks out a property by way of the sensations that it causes, but the word is not an abbreviation for 'the cause of heat sensations'.[17] As for Reid, the fact that heat causes heat sensations is true but not necessary. In some possible worlds, heat causes no sensations at all. In others, heat sensations are caused by effluvia and evil demons. Language users fix the referent of 'heat' by one of heat's accidental identifying features, that of causing heat sensations. To know what heat is or what it is like essentially, scientists must study it empirically. Kripke writes,

> We've discovered eventually that this phenomenon [that causes heat sensations] is in fact molecular motion. When we have discovered this, we've discovered an identification which gives us an essential property of this phenomenon. We have discovered a phenomenon which in all possible worlds will be molecular motion – which could not have failed to be molecular motion, because that's what the phenomenon is. On the other hand, the property by which we identify it originally, that of producing such and such a sensation in us, is not a necessary property but a contingent one.[18]

Fixing the referent of 'heat' by way of our sensations explains nothing about what heat is. Causing sensations is nothing compared to what scientists have discovered by investigating the nature of heat. Once you know what heat is, then you can understand that heat is molecular motion or caloric or phlogiston or whatever describes the essential characteristics of the heat. Kripke goes on,

> In general, science attempts, by investigating basic structural traits, to find the nature, and thus the essence (in the philosophical sense) of the kind. The case of natural phenomena is similar; such theoretical identifications as 'heat is molecular motion' are necessary, though not a priori. The type of property identity used in science seems to be associated with necessity, not with a prioricity, or analyticity: For all bodies x and y, x is hotter than y if and only if x has higher mean molecular kinetic energy than y.[19]

For Kripke, the task of science is to discover the natures, and hence the essential features, of things that are otherwise known only in a relative or contingent way. It is a contingent fact that heat causes heat sensations. You can use that fact to fix the reference for 'heat'. Then, scientific-experimental methods and theoretical genius probe its essential features.[20]

Empirical study is also Reid's prescription for discovering the essential features or objective, physical natures of secondary qualities. Although you obviously know to what your naive concept of heat refers, the cause of your heat sensations, you do not, by your original perception-based conception, know what heat is in itself. Such a study, Reid thinks, requires careful systematic inquiry: 'It is the business of philosophers to investigate, by proper experiments and induction, what heat and cold are in bodies', and, 'Very curious discoveries have been made of the nature of heat, and an ample field of discovery in [the secondary qualities] still remains'.[21] In Hilary Putnam's language, Reid means to analytically specify heat, to say or understand what it is to be heat.[22] Scientists are to search for a conception by which to understand heat distinctly and clearly. Is heat a chemical element or, as Reid suspects and Kripke declares, 'a certain vibration of the parts of a body?'

For Reid, the correct answers to these questions provide direct conceptions of heat because they consider heat as it is in itself, without reference to its causal or other relations. Such answers are obvious instances of a posteriori identifications. Nothing about heat or its identifying marks necessitates a priori its being a certain vibration of the parts of a body. Once you realise that the object of your conception with presentational content 'the cause of heat sensations' is the same thing as the object of your conception with presentational content 'a certain vibration of the parts of a body', this discovery is best expressed by an identification. You say, 'Heat is a certain vibration of the parts of a body.' And this is precisely the sort of a posteriori identification found in Kripke: 'Heat is the motion of molecules'.[23]

Clearly, Reid is calling for empirical identifications. As noted in Chapter 4, he considers several failed attempts to identify the natures of secondary qualities. If identification of a quality with its nature were an a priori matter, then these failings would be either impossible or incredibly rare. Reid mentions, for example, some chemists who conceive of smell as a spiritus rector, 'a kind of soul' that 'flies about in the air in quest of a proper receptacle'.[24] Of course, Reid himself does not endorse such a theory. But these chemists have an inaccurate conception of a physical smell, which just means that they have a conception with smell as its referential content and 'soul that flies about in the air' as its presentational content. The presentational content is inaccurate, yet the referential content makes it clear that the smell is the object of the conception. Reid also mentions a theory of heat on which heat is 'a particular element diffused through nature and accumulated in the heated body'.[25] A misguided scientist, then, might wrongly identify heat as a particular element. These are not failures in perception and not lexical failures of 'heat' either. They are immature speculations, the sorts of momentary failures in empirical research that ultimately lead to successes.

Conclusion

Jackson is right that names given to secondary qualities on the basis of naive perceptions do not appear in scientific theories. However, this does not eliminate them from today's scientific landscape. The language of secondary quality identifications, 'molecular motions', 'vibrating elastic media', 'tiny airborne particles' and 'spectral reflectances', fits readily into today's scientific lexicon. Following Putnam and Kripke, these successful a posteriori identifications, or analytic specifications, imply – indeed, entail – the physicality of secondary qualities. And, as it turns out, by studying physical properties that affect human sense organs, scientists have been following Reid's prescription for discovery.

Secondary qualities don't automatically match up to scientific vocabulary. But they are scientific properties nonetheless. Reid's theory explains why this might be and challenges scientists to identify secondary qualities as they are in themselves. There are several objections to identity theories of secondary qualities, which receive attention in Part III. Before moving on from Reid, Chapter 6 addresses a few remaining controversial questions among interpreters of Reid regarding the proper understanding of the primary/secondary quality distinction.

Notes

1. Reid, *Essays on the Intellectual Powers of Man*, 2.21/235.
2. Reid, *Inquiry into the Human Mind*, 2.2/26.
3. Reid, *Essays on the Intellectual Powers of Man*, 2.21/236 suggests identifying heat as a property of a body via original perceptions.

4. Reid, *Inquiry into the Human Mind*, 2.8/41. Reid tells a similar story about sound in *Inquiry into the Human Mind*, 4.1/50.
5. Jackson, *Perception*, pp. 124–5.
6. Reid, *Inquiry into the Human Mind*, 6.5/86–7.
7. Reid, *Essays on the Intellectual Powers of Man*, 2.17/203.
8. Reid, *Essays on the Intellectual Powers of Man*, 2.17/204.
9. Putnam, 'Meaning and Reference'.
10. Reid, *Inquiry into the Human Mind*, 5.1/54. Compare Reid, *Inquiry into the Human Mind*, 2.8/39 and Reid, *Essays on the Intellectual Powers of Man*, 2.17, where Reid argues that 'smell' and 'heat' are names for qualities of bodies, despite the fact that no one knows their natures.
11. Reid, *Inquiry into the Human Mind*, 5.1/54.
12. Reid, *Inquiry into the Human Mind*, 5.2/57.
13. Kripke, *Naming and Necessity*, p. 119.
14. Kripke, *Naming and Necessity*, pp. 118–19.
15. Kripke, *Naming and Necessity*, p. 117.
16. Kripke, *Naming and Necessity*, pp. 123–5.
17. Kripke, *Naming and Necessity*, pp. 130–6.
18. Kripke, *Naming and Necessity*, p. 133.
19. Kripke, *Naming and Necessity*, p. 138.
20. Putnam describes the scientist's task as discovering the *analytical specification* of the heat or water or other subject of interest. See Putnam, 'Meaning and Reference', p. 708.
21. Reid, *Inquiry into the Human Mind*, 5.1/55; Reid, *Essays on the Intellectual Powers of Man*, 2.17/204.
22. Putnam, 'Meaning and Reference', p. 708.
23. Kripke, *Naming and Necessity*, p. 129.
24. Reid, *Inquiry into the Human Mind*, 2.1/25.
25. Reid, *Inquiry into the Human Mind*, 5.1/55.

6

Understanding Reid's Distinction

Thomas Reid's primary/secondary quality distinction has proven challenging. Reid scholars offer impressively divergent readings from one another. Chapter 4's approach may be unique among these. So some additional precision in interpretation is warranted. If you are impatient to see how Reid's objective theory of secondary qualities handles philosophical scrutiny, try proceeding to Chapter 7 and returning here later. But if your interest is in understanding Reid's writing, exploring his historical significance or grasping the theoretical landscape, Chapter 6 addresses a number of exegetical concerns. These serve to extend or reiterate the interpretation in Chapter 4, minimising the confusion generated by Reid's leaving these details for the reader to sort out.

Epistemic or Metaphysical?

It is not unusual for Reid interpreters to apply Reid's talk of relative secondary quality conceptions to the qualities themselves, positing a metaphysical distinction where there is none. They suggest that primary qualities are intrinsic qualities of bodies whereas secondary qualities are mere relational 'powers to produce certain characteristic sensations in us in normal circumstances'.[1] For example, James Van Cleve holds that secondary qualities are dispositions. From there, he argues, 'If secondary qualities are dispositions, they will differ metaphysically from primary qualities – they will be relational or extrinsic properties . . . whereas primary qualities are intrinsic'.[2] Thus, in addition to whatever distinction arises as a result of original conceptions of perceivable qualities, Van Cleve understands Reid as positing a metaphysical distinction between primary and secondary qualities. Others see Reid's description of secondary qualities as 'that which occasions such a sensation' and his mention of power and virtue, as evidence for the metaphysical distinction.[3] However, these interpreters have misunderstood Reid on this issue. His distinction is purely conceptual.

Van Cleve's argument, which focuses on Reid's secondary qualities as causes of sensations, suggests two ways to take the primary/ secondary quality distinction as metaphysical. First, dispositional/ categorical: secondary qualities are dispositions, abilities, tendencies or powers, best understood in terms of stimulus conditions and manifestations, while primary qualities are categorical. Fragility, for example, is the disposition of an object to shatter when struck. Solubility is the disposition of an object to dissolve in water. Primary qualities, on the other hand, are categorical, like being round or being made of plastic.

Second, intrinsic/extrinsic: secondary qualities are extrinsic or relative and primary qualities intrinsic. Intrinsic properties pertain to the way a thing is in itself. But extrinsic properties concern interaction with the world. The intrinsic property of mass depends only on the quantity of matter composing an object, whereas weight is a matter of the gravitational pull on a massive object due to another body. A trip to the Moon would change your weight, because the Moon's gravitational pull is weaker than Earth's, but not your mass. Likewise, the intrinsic/extrinsic interpretation views secondary qualities as extrinsic due to their supposed dependence on perceivers.

Van Cleve seems to regard the dispositional/categorical distinction as congruent with the intrinsic/extrinsic. This is reasonable, especially since Jennifer McKitrick's impressive argument against intrinsic dispositions.[4] If she is right, then dispositions are necessarily extrinsic and, perhaps, categorical properties are intrinsic. On the other hand, McKitrick's claim opposes a longstanding tradition, including David Armstrong, J. L. Mackie and David Lewis, who accept some dispositions as intrinsic.[5]

Regardless of whether the intrinsic/extrinsic distinction subsumes the dispositional/categorical, neither metaphysical interpretation of Reid's primary and secondary qualities holds up. Consider the intrinsic/extrinsic and dispositional/categorical versions separately, in case McKitrick is wrong about intrinsic dispositions. No passage from Reid compels his readers to think that he endorses a dispositional/categorical or intrinsic/extrinsic version of the primary/secondary quality distinction. And such readings clearly conflict with his assessments of the primary and secondary quality species.

Some may be misled by Reid's references to power and virtue, which sound dispositional. But the jump from power and virtue to disposition is unwarranted. When Reid mentions these in sections 2.8 and 2.9 of his *Inquiry*, he is speaking of one particular secondary quality – the smell of a particular rose.[6] In the scenario Reid imagines, a perceiver smells the rose for the first time, and Reid traces the development of the perceiver's conceptions of the smell, from vague cause to clearly understood effluvia. He attaches the notions of power and virtue to the smell because the hypothetical smeller would have trouble finding better ways to describe it, not because smell is a disposition.

Furthermore, Reid distinguishes between powers and tendencies, which seem to be dispositions:

> Magnetism signifies both the tendency of the iron towards the magnet, and the power of the magnet to produce that tendency . . . The same thing may be said of gravitation, which sometimes signifies the tendency of bodies towards the earth, sometimes the attractive power of the earth, which we conceive as the cause of that tendency.[7]

In Reid's world, powers do not necessarily indicate tendencies but categorical causal bases of tendencies. Moreover, this passage appears in the same section of Reid's *Inquiry* as the references to power and virtue that are taken by others as evidence for secondary qualities as dispositions. If Reid's talk of secondary qualities as virtues and powers were to mark them as dispositional or categorical, it would be the latter. The identification of smells with effluvia makes them categorical, along with heat and sound, which are types of vibrations. On the other hand, not all of Reid's secondary qualities are categorical. Reid explicitly identifies colour as a disposition to reflect certain kinds of light.[8] Secondary qualities include both dispositional and categorical properties.

Primary qualities also overlap the dispositional/categorical divide. Solidity is a disposition to exclude other objects from the solid object's space.[9] Divisibility is, as its suffix suggests, a disposition as well. It is the disposition to divide in some circumstances. Figure and extension, however, are categorical. So the primary qualities, like the secondary qualities, include both dispositional and categorical members. It is impossible to understand him as saying that secondary qualities just are dispositional properties and primary qualities just are categorical properties.

Others who push a metaphysical distinction on Reid's doctrine of primary and secondary qualities claim that primary qualities are intrinsic to the objects that possess them while secondary qualities are extrinsic. This line of thought regards secondary qualities as extrinsic because relational, dependent on or constituted by relations to sensations. For example, Ryan Nichols writes,

> Colors, sounds, tastes and smells are relational properties of objects . . . [They] are powers to produce certain characteristic sensations in us in normal conditions; to ascribe such a quality to an object is not to perceive any intrinsic quality of the object, but is, rather, to perceive that the object bears a certain relation to something else: namely, ourselves.[10]

As Nichols reads Reid, physical objects possess secondary qualities, but as extrinsic relations secondary qualities need perceivers for their existence. Furthermore, he regards Reid's primary qualities as intrinsic.

Relations play a role in Reid's account of secondary qualities, but Nichols speaks too loosely. His interpretation does not hold up in light of Reid's comments on various species of primary and secondary qualities. For Reid, neither primary nor secondary qualities are, by nature, sensation relative. Most likely, the intrinsic/extrinsic interpretation arises from a misreading of Reid's official definitions of primary and secondary qualities. Again, they are,

> There appears to me to be a real foundation for the distinction; and it is this: That our senses give us a direct and a distinct notion of the primary qualities, and inform us what they are in themselves: But of the secondary qualities, our senses give us only a relative and obscure notion. They inform us only, that they are qualities that affect us in a certain manner, that is, produce in us a certain sensation; but as to what they are in themselves, our senses leave us in the dark.[11]
>
> Thus I think it appears, that there is a real foundation for the distinction of primary from secondary qualities; and that they are distinguished by this, that of the primary we have by our senses a direct and distinct notion; but of the secondary only a relative notion, which must, because it is only relative, be obscure; they

are conceived only as the unknown causes or occasions of certain sensations with which we are well acquainted.[12]

According to these passages, secondary qualities are those for which your sense perceptions give you conceptions with presentational contents relative to some sensation caused by the quality. Chapter 4 interprets the description as an analysis of what it is to be a secondary quality. You might formalise the definition,

$$\Box \forall x \, (Secqual \; x \leftrightarrow \exists y \exists z \; Sensation \; y \; Origconcept \; z$$
$$Refcontent \; x, z \; Prescontent \; y, z)$$

That is, by definition, a quality, x, is secondary if and only if there is some sensation, y, and some original conception, z, such that x is the referential content of z and y features in the presentational content of z. Secondary qualities are secondary in virtue of the role that sensations play in original conceptions of them. But this is not an essential feature of any particular quality. It is simply a necessary condition for that quality's being secondary. A same-parents relation to another person is necessary for you to be a sibling, but being a sibling is not among your essential properties. You'd still be you if your little sister hadn't come along. So you originally perceive heat by means of sensation-relative conceptions, but being originally perceived by means of sensation-relative conceptions is not essential to heat.

The intrinsic/extrinsic interpretation likely understands Reid as saying something more like,

$$\forall x \, (Secqual \; x \leftrightarrow \exists! \, y \; Sensation \; y \; \wedge \Box \forall w \; (Causes \; w, y \leftrightarrow w = x))$$

For each secondary quality, x, there is some unique sensation, y, and, by definition, whatever thing, w, causes y, just is x, the secondary quality. The second proposal takes Reid's official definitions to be about secondary quality species – heat, smell and colour – rather than secondary qualities' collective genus. It says that heat, being a secondary quality, is heat in virtue of its causing heat sensations. Colour, being a secondary quality, is colour in virtue of causing colour appearances. The second reading makes sensation-relative conceptions essential to a philosophical analysis of each secondary quality. Secondary qualities depend on sensations for their existence, which makes them both sensation-relative and extrinsic.

On the first reading, sensation-relative original conceptions of a quality are necessary for its being secondary. On the alternative reading, causing sensations is necessary for each quality's existence. The first reading is superior because, unlike the second, it acknowledges Reid's claims that secondary qualities are only accidentally linked to the sensations that they cause. It makes sense of Reid's desire for scientists to investigate the natures of secondary qualities. And it is consistent with the fact that Reid never identifies any perceivable quality as sensation-relative.

If secondary qualities essentially depend on the sensations that they cause, then it is impossible that heat could have been signified by anything other than heat sensations or sound by anything other than sound sensations. But Reid says that they might have been:

> No man can give a reason, why the vibration of a body might not have given the sensation of smelling, and the effluvia of bodies affected our hearing, if it had so pleased our Maker. In like manner, no man can give a reason, why the sensations of smell, or taste, or sound, might not have indicated hardness, as well as that sensation, which, by our constitution, does indicate it.[13]

Just as Kripke links names and their referents contingently, so Reid describes the relations of natural signs and the things that they signify. Sound sensations do not necessarily indicate sounds; they might have signified hardness. If this is possible, then sound is not merely whatever causes sound sensations.

The first reading, unlike the second, allows for scientific learning about the natures of secondary qualities.[14] Again, Reid expects conceptions of secondary qualities to multiply with experience and scientific study, becoming ever more clear and accurate. For example, original conceptions of heat have presentational content 'the cause of that heat sensation'. But without substantial theorising and experimentation, or at least attending science class, you don't know that the cause is in fact molecular motion. For all the unlearned, original conception tells you, it might be caloric, phlogiston or the Greek element of fire. The intrinsic/extrinsic reading regards heat as whatever has a tendency to cause heat sensations. If this is the case, then you know the nature of heat by your original conception. There is no reason to investigate its nature, even if you're curious about its cause.

Consider the implications of the second reading. It makes Reid's definition of secondary quality into a functional one. Each secondary quality is what it is because of the sensations it causes. The implications of this are quite strange. If the sensation of heat is caused by molecular motion in the actual world and by the heat-sensation-inducing actions of a rose's effluvia in another possible world, then the second reading asserts that these two things, molecular motion and the effluvia, are of the same kind. Unified accounts of secondary quality species are impossible, because secondary quality species are world-relative on this reading. At best, there might be disjunctive accounts: heat is molecular motion or effluvia or a disposition to reflect certain kinds of light or the firm adhesion of a thing's parts and so on. Defining secondary qualities in this manner is a little silly, like identifying a play with its title. If Hamlet just were whatever play bears that title, then Hamlet in our world would still be a tragedy about the revenge of a young prince. But in another possible world, it would be a comedy about a small town in Canada, or an opera starring a little Caribbean fish. Why would Reid charge scientists with the investigation of such gerrymandered properties? Surely he wouldn't, and the first reading is preferable to the second.

Finally, Reid never analyses any specific secondary quality as sensation- or otherwise perceiver-relative. For him, sound and heat are types of vibrations,[15] and smell has something to do with tiny particles, independent of anyone's perception of them.[16] You might think that colour is a sensation-relative extrinsic property, because Reid explicitly identifies it as a disposition.[17] However, colour is not specified as a disposition to cause sensations but to reflect certain kinds of light. It may be extrinsic, but it isn't sensation-relative. No secondary quality depends on sensations for its existence. Ultimately, the metaphysical readings of Reid's official definition of the primary/secondary quality distinction create more problems than they solve.

Despite the textual evidence against the metaphysical intrinsic/ extrinsic distinction, some think that other passages in Reid's writing showcase secondary qualities as extrinsic and sensation-relative, in contrast to the intrinsic natures of the primary qualities. Foremost among these, Van Cleve makes his case from this passage, which appears in the same section as Reid's official definition of primary and secondary qualities:

> We may see why the sensations belonging to secondary qualities are an object of our attention, while those which belong to the primary are not.
>   The first are not only signs of the object perceived, but they bear a capital part in the notion we form of it. We conceive it only as that which occasions such a sensation, and therefore cannot reflect upon it without thinking of the sensation which it occasions: We have no other mark whereby to distinguish it. The thought of a secondary quality, therefore, always carries us back to the sensation which it produces. We give the same name to both, and are apt to confound them together.[18]

According to Van Cleve, Reid's description of sensations as 'capital parts' of human notions of secondary qualities commits him to the claim that it is impossible to understand secondary qualities apart from sensations. That is, you cannot conceive of secondary qualities without conceiving of the sensations that they cause. To this claim, Van Cleve adds, as an obvious fact, that you can conceive of things like molecular motion or the vibration of a thing's parts without conceiving of sensations. So these supposed identifications of secondary qualities, like vibrations of bodies or molecular motion, have a property that Reid's secondary qualities lack, namely, the ability to be conceived without reference to a sensation. Therefore, Van Cleve concludes that heat is not identical to molecular motion, or to anything independent of perceivers and their sensations.

Van Cleve's argument is valid. If you can't conceive of heat without conceiving of heat sensations but you can conceive of molecular motion without conceiving of heat sensations, then by Leibniz's Law, heat is not molecular motion. Of course, you can conceive of molecular motion without conceiving of heat sensations. So disagreeing with Van Cleve means offering an alternative reading of the 'capital part' passage or declaring Reid inconsistent. Can you conceive of heat apart from sensations?

The passage makes clear that Reid regards certain conceptions of secondary qualities as sensation-relative. That is, the presentational contents of a least some notions of secondary qualities include references to sensations. But does anything in this passage indicate that all conceptions of secondary qualities include sensation-relative presentational contents? No, why would it? Reid's language here refers to original secondary quality conceptions, not to all conceptions. Again, sensations 'bear a capital part in the notion we form of [a secondary quality]'. 'The' considers only one notion or conception, namely, the one formed by means of unlearned perceptions. But nothing in this sentence prevents you from forming additional conceptions of secondary qualities with presentational contents that make no reference to sensations at all.

Three sentences later, Reid says, 'The thought of a secondary quality, therefore, always carries us back to the sensation'. Although it is not part of Van Cleve's argument, this line may give, the best evidence from the 'capital part' passage for his claim that Reid's secondary quality conceptions must be sensation-relative. It suggests that Reid's intended scope is universal. However, the context does not carry the modal force required to say that all possible thoughts about secondary qualities involve thoughts of sensations.

Why not consider Reid as referring to all human thoughts of secondary qualities that employ perception-based conceptions? That is, Reid seems to mean that original conceptions of secondary qualities involve thoughts of sensations. As the introductory lines of this passage show, Reid is explaining why perceivers typically notice secondary quality sensations but not primary quality sensations. His answer is that original, perception-based conceptions of secondary qualities relate those qualities to the sensations that they cause, but not so for conceptions of primary qualities. The presentational contents of original conceptions of secondary qualities infect thoughts with conceptions of sensations. And this is always true, so long as you consider them by means of original conceptions. Your perception-based conceptions of primary qualities, by contrast, are not sensation-relative, so you don't tend to think about sensations when considering primary qualities by your original conceptions. It is difficult, Reid says, to attend to primary quality sensations, unless they are pleasant or painful.[19] This understanding captures Reid's concern for the role of sensations in secondary quality conceptions without violating his other commitments. Nothing in the 'capital part' passage demonstrates that thoughts about secondary qualities must involve thoughts about sensations.

Is there evidence that Reid allows secondary qualities to be considered apart from sensations? Yes. On the same page as the 'capital part' statement, Reid identifies a variety of secondary qualities without making reference to sensations:

> The nature of secondary qualities is a proper subject of philosophical disquisition; and in this philosophy has made some progress. It has been discovered, that the sensation of smell is occasioned by the effluvia of bodies; that of sound by their vibration. The disposition of bodies to reflect a particular kind of light occasions the sensations of colour. Very curious discoveries have been made of the nature of heat, and an ample field of discovery in these subjects remains.[20]

Here Reid outlines conceptions of various secondary qualities' natures. He identifies them with effluvia of bodies, vibrations and spectral reflections. None of these things is sensation-relative. So, standing Van Cleve's argument on its head, consider the following counter-argument: sound is the vibration of a body and you can conceive of the vibration of a body without conceiving of sound sensations; therefore, you can conceive of sound without conceiving of sound sensations. Reid's identification of sound with vibrations is much more explicit than his supposed claim that it is impossible to conceive of secondary qualities without conceiving of secondary quality sensations in the 'capital part' text. It's doubtful that Reid understands individual secondary qualities as sensation-relative by nature.

A second reason to avoid the claim that you can't conceive of secondary qualities except in a sensation-relative manner is that you obviously can. After reading this sentence, you'll have a conception of heat with presentational content 'the quality Reid mentions on pages 54 and 55 of the *Inquiry*'. I have another with presentational content 'the quality of a particular lawn mower muffler that burned my left hand'. Neither of these conceptions is sensation-relative. The first is relative to a page in a book, and the second is relative to an event. You could have these conceptions of heat, even if you lacked the ability to perceive heat by means of heat sensations. You would just think of heat as Reid's topic on pages 54 and 55 of the *Inquiry* or the cause of my injury. So it is possible to conceive of secondary qualities without conceiving of sensations. If Reid thinks otherwise, he has erred.

As observed in previous sections, Reid's distinction between primary and secondary qualities depends on conceptions formed by means of perception. Perception-based conceptions of secondary qualities comprehend those qualities by way of sensations that they cause. But there is no reason to suppose from this that secondary qualities themselves depend on the sensations they cause. A relative conception of a thing does not make the thing itself relative. In fact, you can form relative conceptions of nearly anything. Secondary qualities, like the vibration of a bell and the molecular motion of a cup of coffee, are what they are no matter your sensations.

There is no reason to posit a metaphysical distinction in Reid's theory of primary and secondary qualities in addition to the conceptual distinction introduced in Chapter 4. A dispositional/categorical distinction will not work, because some secondary qualities are categorical and some primary qualities are dispositions. An intrinsic/extrinsic distinction, where secondary qualities are relative to sensations, will not work, because the causal connections between secondary qualities and their sensations are accidental and because such a view makes the scientific investigation of secondary qualities futile. Finally, the fact that original conceptions of secondary qualities are relative to sensations does not limit types of conceptions one can have of them or dictate what they must be. For Reid, there is no metaphysical distinction between primary and secondary qualities. The distinction is epistemic or conception-based, and therefore it is a feature of human minds, not of the primary and secondary qualities themselves.

Conceptual, Phenomenological, or Both?

Some seem to find a perceiver-relative primary/secondary quality distinction in Reid besides the conceptual one expounded in Chapter 4. They claim a distinction based on the phenomenology of primary and secondary quality perceptions. One version of this interpretation says that perceptions of primary and secondary qualities differ with regard to awareness of accompanying sensations. It suggests that, according to Reid, you always notice sensations when perceiving secondary qualities but never when perceiving primaries. A more radical version says that you always have sensations when perceiving secondary qualities and never when perceiving primary qualities. Both of these readings misunderstand Reid.

The more radical phenomenological distinction focuses on Reid's characterisation of primary quality perceptions as direct and immediate. Ryan Nichols, for example, takes these words, 'direct' and 'distinct', to describe differing causal processes in perceptions of primary and secondary qualities, not differing presentational contents. He writes,

> While directness concerns the formation of our concepts of qualities, Reid also argues that the contents of our conceptions of primary and secondary qualities differ in crucial respects. This marks the second means by which Reid draws his distinction.[21]

Nichols says that, by casting primary quality perceptions as direct and immediate, Reid explains something about how primary quality conceptions are formed, namely, without causal cognitive mediation. There are no characteristic psychological events preceding primary quality perceptions, and no sensations in particular. Nichols offers the formal description,

> P's notion of quality Q is a notion of a primary quality only if P apprehends Q, and no intermediary is necessarily apprehended in the process.[22]

When perceiving primary qualities, Nichols thinks, your mind address the quality without pausing to consider, or perhaps even to experience, sensations. Moreover, he argues, Reid expressly identifies the perceivable property of visible figure, perhaps best understood as a two-dimensional geometric projection of an object's silhouette on to a single point, a primary quality lacking a corresponding sensation.[23] So this proposal boasts support from Reid's notes on primary quality species as well as those on primary qualities in general.

It is a mistake to take direct and immediate in a causal or chronological sense. Contra Nichols, direct and immediate conceptions carry direct and immediate presentational contents. They don't circumvent perceptual processes. Reid describes notions of primary qualities as direct in his official definition:

> There appears to be a real foundation for the distinction; and it is this – that our senses give us a direct and distinct notion of the primary qualities. . . . But of the secondary qualities, our senses give us only a relative and obscure notion.[24]

In this passage, Nichols takes Reid to describe how you form conceptions of primary qualities: directly rather than by some Rube Goldberg process. But, grammatically, it seems to be the notion of the primary quality that is direct, not the manner in which the notion is given. If Reid means to say that you form conceptions in a causally or chronologically direct way, he should use 'direct' adverbially, as in 'our senses directly give us'. What, again, is a direct notion? One that understands its object directly, rather than through causal or other relations. 'Direct' indicates a type of presentational content, not a causal origin.

On the other hand, Nichols is right that Reid describes perceptions as immediately formed. The term appears in the earlier-mentioned 'capital part' passage, about why perceivers typically do not attend to primary quality sensations. There, Reid says, 'When a primary quality is perceived, the sensation immediately leads our thought to the quality signified by it, and is itself forgot.'[25] Reid describes the perceptual process as immediate to indicate that the conception and belief come from a distinct faculty, not reason, not memory, not testimony. His point is that, because you don't reason from sensations to qualities and because sensations don't feature in the presentational contents of original conceptions of primary qualities, you have little opportunity to consider the sensations that primary qualities cause. So even though original conceptions of primary qualities form when sensations trigger them, most perceivers have not paused to consider those sensations. You form your conception of hardness when you experience sensations caused by the hardness of some object, but the conception you form of the hardness has the presentational content 'the firm adherence of the parts of a body'. You often don't consider the hardness in terms of the sensation it causes. So your conception forms just as with a secondary quality. The quality causes a sensation that triggers a conception and belief in the quality. But you consider the quality in itself, not relative to the triggering sensation.

Visible figure, the phenomenological reading's paradigm case, doesn't help the radical interpretation either. First, even if it is perceivable without sensations, such a feature uniquely distinguishes it from all other perceivable properties, rather than demarcating primary and secondary qualities. Second, Reid neither says nor suggests that visible figure is a primary quality. One interpreter assigns it to the primary qualities on the basis of its mind-independence.[26] But mind-independence can hardly make for primary-ness. In a letter to Lord Kames, Reid says the same of the taste of sugar, a secondary quality.[27] Third, Reid does not say that visible figure is perceivable without sensations. He merely says that there is no characteristic sensation associated with it. It is consistent with Reid's text that visible figure is perceived in virtue of a variety of sensation types, visual and tactile.[28] There is no reason to construe Reid's comments concerning visible figure as the basis for a phenomenological primary/secondary quality distinction. Nor should one suppose that human perceptions happen without sensations.

A milder phenomenological distinction says that you cannot be aware of primary quality sensations, but you do notice secondary quality sensations. In this vein, Keith Lehrer remarks,

> The distinction that Reid has drawn is, in fact, both phenomenologically and conceptually based. As he insists, in the case of primary qualities, we hardly notice the character of the sensation but immediately take the sensation as the sign of the quality.[29]

Tony Pitson, after considering the conceptual distinction, also writes,

> Another [contrast between primary and secondary qualities] is at the level of the sensations or experiences involved . . . The sensations which belong to the secondary qualities are in general objects of attention, but those which belong to the primary qualities are not. This makes it possible for the sensations involved to form part of the very conception of secondary qualities. In the case of primary qualities, sensation leads immediately to the thought of the quality signified and we have no occasion to attend to it.[30]

It is not clear how seriously Pitson takes the phenomenological distinction since he qualifies his statement with 'in general'. But since he remarks on it alongside the conceptual distinction, let's treat him as if he understands Reid as offering a phenomenological basis for the primary/secondary distinction.

Both of these interpreters support their positions with this passage, which you may recall from the last section:

> But having a clear and distinct conception of primary qualities, we have no need when we think of them to recall their sensations. When a primary quality is perceived, the sensation immediately leads our thought to the quality signified by it, and is itself forgot. We have no occasion afterwards to reflect upon it; and so we come

to be as little acquainted with it, as if we had never felt it. This is the case with the sensations of all primary qualities, when they are not so painful or pleasant as to draw our attention.[31]

In accordance with Pitson and Lehrer's remarks, the words do suggest a way in which sensations differ with regard to each quality type. But adding this point into Reid's account of the distinction between primary and secondary qualities is misguided. In this passage, Reid is offering an error theory for how other philosophers have missed the fact that primary quality sensations, type-3 natural signs, do not resemble external objects. His answer is that they did not direct their attention to these sensations because there is rarely a need to do so.

Attention, according to Reid, is the operation of the mind by which one forms distinct notions of things or discovers their properties.[32] It is how you learn about things. Although you can voluntarily direct your attention, odds are that you attend to primary qualities rather than the sensations they cause. Just as there is usually no need for an able reader to study the spelling of a word or the font in which it is written, so primary quality sensations communicate to you so readily that you habitually ignore them. However, this habit can be broken.

Lehrer's and Pitson's proof text does not indicate that the phenomenological difference makes primary qualities primary or secondary qualities secondary. It merely mentions a cognitive trend. Someone might distinguish primary qualities from secondary in other accidental ways, perhaps by noting that primary qualities are those addressed first in Chapter 4 of this book or used most in scientific explanations. This distinction would be accurate but nothing that interests Reid. The fact that most perceivers tend to attend to secondary quality sensations and not primary quality sensations is an interesting observation about human behaviour but not a philosophical doctrine.

Another problem with the mild phenomenological interpretation is that, if it is taken too strictly, it generates an inconsistency in Reid's text. A strict interpretation would say that perceivers never attend to primary quality sensations and always to secondary quality sensations. But there are clear counterexamples to this. For example, one may perceive the secondary quality of colour without attending to its accompanying sensation. Reid explains,

> In [visually] feeling a coloured body, the sensation is indifferent, and draws no attention. The quality in the body, which we call its colour, is the only object of attention; and therefore we speak of it, as if it were perceived, and not felt.
>
> There are some sensations, which, though they are very often felt, are never attended to, nor reflected upon. We have no conception of them; and therefore, in language, there is neither any name for them, nor any form of speech that supposes their existence. Such are the sensations of colour, and of all primary qualities; and therefore those qualities are said to be perceived, but not to be felt.[33]

Thus, colour stands as a glaring counterexample to the idea that secondary qualities can't be perceived without attending to their corresponding visual sensations.

Not only do the phenomenological distinctions want of textual evidence, but also Reid can't have them, since these readings would undermine his case against the Way of Ideas. He insists that conceptions of primary qualities are triggered by sensations that do not resemble the qualities that they signify. To make such a claim, Reid must attend to his primary quality sensations. He acknowledges that the task as difficult.[34] But if inattention to primary quality sensations were anything more than a habit, Reid would have no way to discover the natures of primary quality sensations or understand that they do not resemble primary qualities. Without this information, Reid's natural sign theory could not extend to the primary qualities, and he could not account for sensations in his general account of perception.

In fact, sensations that prompt perceptions of primary qualities are the very thing that Reid believes himself to have studied. He claims,

> Let a man press his hand against a hard body, and let him attend to the sensation he feels, excluding from his thought every thing external, even the body that is the cause of his feeling. This abstraction indeed is difficult, and seems to have been little, if at all, practised: But it is not impossible, and it is evidently the only way to understand the nature of the sensation. A due attention to this sensation will satisfy him, that it is no more like hardness in a body, than the sensation of sound is like vibration in the sounding body.[35]

Again, in an abstract of the *Inquiry* he sent to Revd Dr Blair, Reid claims to have broken the habit of ignoring primary quality sensations and offers an explanation of how he accomplished this:

It is extremely difficult to attend to the sensations that are neither pleasant nor painfull, such as those we have when we feel a body hard or soft, rough or smooth of this or that figure. . . .

I flatter myself that by much pains and practice I have overcome this difficulty in some degree. When I present an Object to any of my Senses in order to attend to the impression that is made upon the mind, I endeavour to withdraw my thoughts from every thing external, to turn them inward and consider purely what I feel. I suppose every external existence annihilated, every impression and thought I ever had before quite obliterated, and that I begin a new Scene of Existence with this single Impression. What is it like? To what is it like? I view it narrowly on every side and resist every thought that would divert my Attention untill I be well acquainted with it, and able to make it an object of thought.

When I had acquired the power of thus attending to my Sensations [when perceiving primary qualities], I was soon perswaded that I had never made them objects of thought before & that those Sensations which I had felt every day, perhaps every hour of my life, had notwithstanding been as much unknown to me as if I had never felt them, because I had never given any attention to them.[36]

Habits of attending to secondary quality sensations and ignoring primary quality sensations may be strong, they are not invulnerable. Human minds may attend signs for either primary or secondary qualities. The phenomenological distinction between primary and secondary qualities is flimsy. It can't be essential, since Reid believes he has overcome it. Moreover, his case against the Way of Ideas depends on his own and his reader's ability to form two conceptions, one of a primary quality and another of the sensation that accompanies the perception of that quality. This, again, would be impossible if neither Reid nor his readers ever made primary quality sensations objects of their attention.

When trying to understand Reid's text, introducing a phenomenological difference between primary and secondary quality perceptions creates needless complications. Reid's primary/secondary distinction hinges on the presentational contents of original perceptions. Original perceptions of primary qualities consider those qualities as they are in themselves, hardness as the firm adherence of the parts of a body. Original conceptions of secondary qualities consider them in relation to sensations. For this reason, human perceivers tend to pay more attention to secondary quality sensations and to ignore primary quality

sensations. But this is an accident, not part of the primary/
secondary distinction itself.

Both Dispositional and Causal?

Jennifer McKitrick finds another potential problem with Reid's
account of primary and secondary qualities and points her reader
towards the solution.[37] Reid explicitly identifies certain perceivable qualities – solidity, divisibility and colour – as dispositions.
But each of these qualities also causes of some sensation. So Reid
seems to invoke causally potent dispositions. But Prior, Pargetter
and Jackson have argued forcefully that dispositions are, strictly
speaking, non-causal.[38] This is why a disposition requires a causal
base, some categorical explanation for its potency. The relevant
causal quality of a glass, when considering its fragility, is not its
disposition to shatter when struck. Rather the cause of its breaking
has something to do with its molecular structure and other micro-
physical features, features that do things. These form a causal base
upon which the disposition supervenes. Thus, a critic of Reid may
argue that, if perceivable properties cause sensations, then they
must be causally potent. So why not suppose them causal bases
rather than dispositions? Why not reject Reid's theory for the one
that can explain how divisibility, solidity and colour can cause
sensations?

McKitrick answers that Reid's causes work a little differently
from Prior's, Pargetter's and Jackson's. This is what Reid calls
'cause in the popular sense'. Consider his account of the causal connection between a smell of the rose and the sensation it occasions:

> This [rose that occasions the sensation], when found, we call the
> cause of it; not in a strict and philosophical sense, as if the feeling
> were really effected or produced by that cause, but in a popular
> sense: for the mind is satisfied, if there is a constant conjunction
> between them; and such causes are in reality nothing else but laws
> of nature.[39]

'Constant conjunction' recalls David Hume's account of causation, and 'laws of nature' seem to anticipate deductive-nomological
explanations in science.[40] Reid even considers laws of nature, like
the universal disposition of massive objects to undergo gravitational

attraction, as true causes.[41] If this is the case, then there is no inconsistency in appending the notion of causation to dispositional qualities, since dispositions are constantly conjoined in causal chains. There is a natural law, for instance, that relates a thing's disposition to reflect certain kinds of light to colour sensations in perceivers. Likewise, the disposition to divide in certain circumstances is coupled with human perceptions of that disposition. Reid uses a broad, natural-law-oriented sense of causation, so this potential problem with his account is diffused.

Conclusion to Part II

This chapter and the two preceding it have shown that Reid's distinction between primary and secondary qualities rests entirely on the presentational contents of conceptions formed of those qualities by means of original sense perceptions, and it has eliminated alternative readings that wrongly ascribe ontological or phenomenological aspects to that distinction. The only important primary/secondary demarcation is this: perceptions originally yield direct and distinct conceptions of primary qualities and relative and obscure conceptions of secondary qualities. The key similarity is that both primary and secondary qualities are possessed by physical objects, even if some of these qualities are identifiable with other properties. Reid's theory of secondary qualities does exactly what it must to reject the Non-Physicality Thesis (NPT) from the Problem of Secondary Qualities.

In Part I, NPT found support in a claim that secondary qualities do not feature in contemporary scientific theories. Science alone can account for the causal origins of sensations, so secondary qualities turn out to be superfluous. Human sensations would be the same even if physical objects possessed no secondary qualities, so sensations don't count as evidence for the physicality of secondary qualities. According to the Problem of Secondary Qualities, secondary qualities probably aren't possessed by physical objects and would make no difference if they were.

Reid's theory offers promising grounds for rejecting NPT and its so-called scientific support. Perhaps secondary qualities are in fact scientific properties, even if they don't appear in contemporary scientific theories, at least not by their traditional names. Perhaps human perceivers often fail to properly identify secondary qualities with scientific properties. Reid makes an identity theory of secondary qualities plausible by separating sensations and perceptions as essentially independent but accidentally and causally linked operations of the mind. Sensations, he says, serve as natural signs for secondary qualities, but they themselves are not secondary qualities. Secondary quality sensations serve as type-1 natural signs, so they do not automatically convey to their subjects anything about the natures of the things they signify. That is, in original perceptions of secondary qualities, the conceptions by which you apprehend those qualities are relative and obscure. You learn that secondary qualities cause certain sensations but, without experience and rigorous scientific study, you do not understand them in themselves.

As far as original perceptions alone are concerned, secondary qualities could be, epistemically speaking, anything: vibrations of bodies, elements diffused throughout nature, spiritus rectors that fly about the air, effluvia or something entirely different. All anyone knows is that secondary qualities cause sensations. In other words, original perceptions of secondary qualities open a door to speculating about the natures of those qualities. If physical causes must be scientific, then so must secondary qualities. But it is still up to science to discover their various types and the means by which they affect sense organs.

Secondary qualities may sound a lot like primary qualities. Metaphysically speaking, that is true. Secondary qualities on Reid's theory are causally potent, physical, real properties of bodies, just like the primaries. However, according to Reid, the primary qualities differ from secondary qualities because human perceivers have no need to scientifically investigate the natures of the primary qualities. They are originally perceived by means of type-3 natural signs, so everyday observers immediately and automatically understand what they are and form direct and distinct notions of them. So they show up in the mathematical sciences, where reasoning about them is important. It's not that scientists should investigate or study their natures. It's that original perceptions of primary qualities provide the conceptual resources for understanding things in the physical world. So making discoveries about the natures of secondary and other qualities means coming to understand them by means of primary quality notions: heat as the vibration of a thing's minute parts, sound as the vibration of an entire body, colour as a disposition to reflect certain kinds of light. Primary qualities enjoy a privileged conceptual status because identifications of secondary qualities come in terms of primary quality conceptions. But the primary qualities themselves are no more causal or scientific than the secondaries.

Identifying secondary qualities as scientific properties means that secondary qualities are not causally superfluous. Understanding them as possessed by physical objects does not suggest an overdetermined causal process from physical object to sense organ to sensation. Rather, it means that the scientific properties that cause secondary quality sensations are secondary qualities. And it is wrong to say that secondary qualities do not feature in scientific theories. They do, but their appearances can be covert. It is the scientist's task to discover and explain where and how secondary qualities fit into the scientific image of the world. Moreover, NPT, which says that secondary qualities are not possessed by physical objects, is false. If the scientific properties that cause sense experiences are possessed by physical objects, then so are secondary qualities.

Part III entertains and responds to objections. For these, two features of Reid's account are especially important. First, sensations are contingent accompaniments to perceptions, not parts of the perceptions themselves. Second, perceptual conceptions of secondary qualities are relative and obscure. They do not commit to very much concerning the natures of the secondary qualities, just to their existence and causal efficacy. Most of the objections that have been brought against the identity theory of secondary qualities rashly assume that one of these points is false.

Notes

1. Nichols and Yaffe, 'Thomas Reid'.
2. Van Cleve, 'Reid on the Real Foundation of the Primary–Secondary Quality Distinction', p. 280.
3. McKitrick, 'Reid's Foundation for the Primary/Secondary Quality Distinction', p. 74; Wolterstorff, *Thomas Reid and the Story of Epistemology*, pp. 112–13.
4. McKitrick, 'A Case for Extrinsic Dispositions'.
5. This tradition includes Armstrong, *Belief, Truth and Knowledge*; Mackie, 'Dispositions, Grounds and Causes'; Lewis, 'Finkish Dispositions'; and Menzies, 'Critical Notices of *Nature's Metaphysics* by Alexander Bird'.
6. Reid, *Essays on the Intellectual Powers of Man*, 2.17/204; Reid, *Inquiry into the Human Mind*, 2.8/38, 2.9/40, 6.4/86, 6.5/87.

7. Reid, *Inquiry into the Human Mind*, 2.9/41–2.
8. Reid, *Essays on the Intellectual Powers of Man*, 2.17/204.
9. Reid, *Essays on the Intellectual Powers of Man*, 2.17/201.
10. Nichols and Yaffe, 'Thomas Reid'.
11. Reid, *Essays on the Intellectual Powers of Man*, 2.17/201.
12. Reid, *Essays on the Intellectual Powers of Man*, 2.17/202.
13. Reid, *Inquiry into the Human Mind*, 5.2/57.
14. Reid, *Essays on the Intellectual Powers of Man*, 2.17/204.
15. Reid, *Essays on the Intellectual Powers of Man*, 2.17/20; Reid, *Inquiry into the Human Mind*, 5.1/55.
16. Reid, *Inquiry into the Human Mind*, 2.9/43.
17. Reid, *Inquiry into the Human Mind*, 6.5/57.
18. Van Cleve, 'Reid's Theory of Perception', pp. 110–11; Van Cleve, 'Reid on the Real Foundation of the Primary–Secondary Quality Distinction', pp. 281–2. See also Reid, *Essays on the Intellectual Powers of Man*, 2.17/204.
19. Reid, *Essays on the Intellectual Powers of Man*, 2.17/204–5.
20. Reid, *Essays on the Intellectual Powers of Man*, 2.17/204.
21. Nichols, *Thomas Reid's Theory of Perception*, p. 167.
22. Nichols, *Thomas Reid's Theory of Perception*, p. 167.
23. Nichols, *Thomas Reid's Theory of Perception*, p. 169. There is a nice exchange on this issue in Yaffe, 'Reid on the Perception of Visible Figure'; Falkenstein and Grandi, 'The Role of Material Impressions in Reid's Theory of Vision'; and Yaffe, 'The Office of an Introspectible Sensation'. See also Reid, *Inquiry into the Human Mind*, 6.8/101.
24. Reid, *Essays on the Intellectual Powers of Man*, 2.17/201; Nichols, *Thomas Reid's Theory of Perception*, p. 166.
25. Reid, *Essays on the Intellectual Powers of Man*, 2.17/204; Nichols, *Thomas Reid's Theory of Perception*, p. 166.
26. Nichols, *Thomas Reid's Theory of Perception*, p. 117.
27. Reid, *The Correspondence of Thomas Reid*, 20 December 1778, 61/115.
28. Reid, *Inquiry into the Human Mind*, 6.7/96.
29. Lehrer, 'Reid on Primary and Secondary Qualities', p. 189.
30. Pitson, 'Reid on Primary and Secondary Qualities', p. 20.
31. Reid, *Essays on the Intellectual Powers of Man*, 2.17/204.
32. Reid, *Essays on the Active Powers of Man*, 2.3/60.
33. Reid, *Essays on the Intellectual Powers of Man*, 2.18/212; Pitson, 'Reid on Primary and Secondary Qualities', pp. 20–1.
34. Reid, *Essays on the Intellectual Powers of Man*, 2.17/204–5.
35. Reid, *Essays on the Intellectual Powers of Man*, 2.17/209; Reid, *Inquiry into the Human Mind*, 5.4/60–1.
36. Reid, *Inquiry into the Human Mind*, Abstract/258–9.
37. McKitrick, Jennifer, 'Reid's Foundation for the Primary/Secondary Quality Distinction'.

38. Prior et al., 'Three Theses about Dispositions'.
39. Reid, *Inquiry into the Human Mind*, 2.9/40.
40. Callergard, *An Essay on Thomas Reid's Philosophy of Science*, pp. 88–90.
41. Reid, *Essays on the Intellectual Powers of Man*, 2.6/103.

PART III

Objections to Reid's Theory of Secondary Qualities and Replies

The first part of this book introduced the Problem of Secondary Qualities and its threat to Direct Realism. It argued that the best answer to the Problem involves a rejection of its Non-Physicality Thesis (NPT), that physical objects do not possess secondary qualities. This means endorsing an account of secondary qualities as identical to certain physical, scientific properties. To this end, Part II considered Thomas Reid's theory of secondary qualities and found it an adequate alternative to NPT. Reid says that secondary qualities happen to cause secondary quality sensations, may be understood as causes of human sensations, and are best understood in physical, primary quality terms. Furthermore, development of such advanced understandings is the prevue of scientific investigation. By perception alone, you can understand secondary qualities only as unknown causes of sensations.

Although Reid's theory allows the rejection of NPT, and thereby the Problem of Secondary Qualities, many have thought it impossible to accurately describe secondary qualities in scientific terms. They think that Reid's programme is a dead end. These final chapters detail eight prominent objections to the identity theory and offer replies to each. None of the eight is specifically directed against Reid, perhaps because his account of primary and secondary qualities has been widely misunderstood and so ignored as a viable solution to the Problem of Secondary Qualities. However, answering them is important because they purport to demonstrate the futility of any programme to identify secondary qualities with scientific properties.

The first five objections focus on the absence of correlations between secondary quality sensations and known scientific properties. They say that since scientific property types do not neatly correspond with sensation types, secondary qualities cannot be physical or scientific. The sixth says that the identity theory of secondary qualities cannot fulfil a necessary requirement – a priori connections between secondary qualities as humans perceive them and the scientific properties with which one might identify them. That is, the identity theory fails because there are no a priori connections between explanans and explanandum. The seventh deals with the role of phenomenology in human knowledge of secondary qualities. It says that sensations are essential for understanding the natures of secondary qualities but not for scientific properties. Since secondary qualities have a feature that scientific properties lack, Leibniz's Law bars the possibility of their identification. The last objection, based on the ante-Reidian resemblance theory of perception, says that mental secondary quality ideas do not resemble objective properties of physical things, as they must to connect perceivers to the physical world. Replies to the first seven objections are my own thoughts and applications of Reid's theory, but treatment of the final objection is more exegetical, analysing the role that primary and secondary qualities play in Reid's attack on the Way of Ideas.

7

Scientific Objections

According to John Locke, the traditional authority on primary and secondary qualities, the effects of secondary qualities wholly depend the operations of the primary.[1] This is compatible with Reid's identity theory. However, Locke only hesitantly ascribes secondary qualities to physical objects, regarding them as 'nothing in the objects themselves', 'powers barely', 'mere powers' and 'imputed' powers.[2] Why does Lock write as though these properties are unreal? Because qualities of physical objects and secondary quality sense experiences do not exhibit strong correlations or, as Descartes calls it, 'exact correspondence'.[3] Locke offers several examples of secondary quality ideas changing without variation in supposed physical objects of perception: looking at coloured objects in the light and in the dark, tasting crushed and uncrushed almonds, and placing hands of different temperatures into lukewarm water.[4] This lack of correlation between objective property and subjective sensation or idea underlies the most prolific and important series of objections to the identity theory of secondary qualities, which is why this chapter considers five non-correspondence objections.

These five are variations on one theme: there is no single-valued function from perceiver-independent physical property to sensation type, and none from sensation type to property type. Four of the five involve non-correspondence in the first direction, from property to sensation. They make it apparent that one physical property in fact causes many different sensations. The fifth, the case of metamerism, demonstrates non-correspondence the other way. Each of several sensations is caused by many different physical properties. However, this non-correspondence, regardless of direction, does not warrant the rejection of the identity theory of secondary qualities. Those who have argued in this manner have assumed either that sensations are essential constituents of perceptions or that secondary quality perceptions commit human perceivers to more than they in fact do.

Others have tried to solve the non-correspondence objection in other ways. Michael Huemer, for example, when faced with the lack of correspondence between colour sensations and scientific physical properties, like spectral reflectance, offers a promissory note, that scientists will find a correspondence once they learn to think of the scientific properties in the relevant categories, whatever those happen to be. That is, no one has really discovered that there is no correlation between secondary qualities and scientific properties, since not every possible correlation theory has been tried.[5] Another approach has been to standardise viewing conditions, so that there is an appropriate sensation linked to each secondary quality. These standard viewing conditions may be specified in a number of ways. Michael Tye merely mentions 'normal viewing conditions' for colours, which seem to include certain contrast effects, viewing distances, lighting conditions and so on.[6] Fred Dretske indexes veridical perceptions of secondary qualities to certain features of evolutionary history.[7] Finally, some have considered that secondary qualities might be relative, although not to perceivers. For example, Tye considers the effects of contrasts within the visual field, and then he entertains and rejects an account of colour on which colours depend relationally on one another.[8] On this theory, you would see, by means of secondary quality perceptions, relations among physical properties, even if you don't see the physical properties themselves. The last section considers these alternative solutions only in so far as they highlight the difficulty of answering the objections presented here, particularly with regard to colour.

Locke and Berkeley's Water Bowl

Consider the heat-related case of the lukewarm water bowl. Although this scenario appears first in Locke,[9] Berkeley offers the most celebrated articulation in his *Three Dialogues*:

> Phil. Those bodies, therefore, upon whose application to our own, we perceive a moderate degree of heat, must be concluded to have a moderate degree of heat or warmth in them; and those, upon whose application we feel a like degree of cold, must be thought to have cold in them.

Hyl. They must.

Phil. Can any doctrine be true that necessarily leads a man into an absurdity?

Hyl. Without doubt it cannot.

Phil. Is it not an absurdity to think that the same thing should be at the same time both cold and warm?

Hyl. It is.

Phil. Suppose now one of your hands hot, and the other cold, and that they are both at once put into the same vessel of water, in an intermediate state; will not the water seem cold to one hand, and warm to the other?

Hyl. It will.

Phil. Ought we not therefore, by your principles, to conclude it is really both cold and warm at the same time, that is, according to your own concession, to believe an absurdity?

Hyl. I confess it seems so.[10]

Berkeley seems to have found a case in which you simultaneously perceive contrary properties, warm and cold, in a single physical object. Unless your perceptual faculties are frightfully prone to error, both perceptions are veridical. Berkeley reasons along the following lines: if you perceive the water as warm, then the water is warm.; if you perceive the water as cold, then the water is cold; you perceive the water as both warm and cold. But this is a problem, because warm and cold are contraries, mutually exclusive in a single body or bowl of water, unless they're mental properties. Maybe a bowl of water can be warm to your left hand and simultaneously cold to your right, but it cannot be absolutely both warm and cold. Thus, warm and cold are mental or subjective and therefore not identifiable in the manner proposed by Reid.

Because of his place in history, the water bowl case is the only version of the non-correspondence argument to which Reid responds or can respond directly. He rejects the premise that you perceive the water as both warm and cold, since it relies on the all-too-common confusion between sensation and perception. Ironically, Reid sees the water bowl scenario as an occasion for noticing this distinction. He says,

Heat signifies a sensation, and cold a contrary one. But heat likewise signifies a quality or state of bodies, which hath no contrary, but different degrees. When a man feels the same water hot to one had, and cold to the other, this gives him occasion to distinguish between the feeling, and the heat of the body; and although he knows that the sensations are contrary, he does not imagine that the body can have contrary qualities at the same time.[11]

According to Reid, you as an experienced perceiver do not for a moment suspect that the water has contrary qualities, as Berkeley suggests. Although you experience two sensation types, you form only one belief about the physical heat of the water. There are no contradictions or attributions of contrary qualities. Rather, the singular heat of the water is signified by two somewhat different sensations.

On Reid's account of secondary quality perceptions, the presentational content of an unlearned perception of heat is 'whatever caused *that* sensation'. So, although you might say that you perceive a quality causing a warm sensation and one causing a cold sensation, questions concerning the natures of these causes are still open. Reid would not tolerate full-blown commitments to perceiving contrary qualities. Not only would that say something about the existence and causal features of the perceived thermal qualities, but it commits you to a particularly detailed conception of the qualities' measures, warm or cold, as well as to a perfect correlation between those measures and your sensations. Such complex commitments simply do not occur in naive perceptions. Understanding heat in terms of degrees and investigating potential correlations between those degrees and your sensations are tasks that require extensive experience and perhaps scientific methodology. You may experience a warm-type sensation upon touching the water, but this does not mean that you perceive warmth in the water.

Reid's response to the water bowl case highlights an important exegetical point. Namely, Reid's description of secondary qualities as unknown causes of known sensations cannot be an endorsement of any correlation between sensation types and physical property types. When Reid defines secondary qualities as unknown causes of known sensations, he does not mean to say this:

> Human perceivers do not, by perception alone, know what causes our various sensations, but they can assign names to the secondary quality types based on those sensations. For example, objects that cause hot-type sensations they call 'hot', sounds that cause concert-pitch-A-type sensations they call 'concert-pitch A', and spectral reflectances that cause green-type sensations they call 'green'. Thus, secondary quality sensations inform perceivers what type of secondary qualities they perceive – hot, concert-pitch A or green – even if the physical natures of those qualities are hidden in perception. Eventually, science may discover that hotness is a high average kinetic energy among molecules, A is 440 Hz vibration, and green is such-and-such spectral reflectance.

If this interpretation were correct, then Reid couldn't deny that you perceive hot and cold properties since you experience warm- and cold-type sensations. But Reid does not take this line. Rather, he claims that, despite contrary sensations and the dissimilar states of your hands, you rightly perceive only one quality in the water. If he is consistent at all, then Reid cannot think that you, even as a matter of common sense, posit or assume an invertible function from sensation types to secondary quality types, as this misinterpretation suggests.

In light of this difficulty, perhaps it's worth recalling that objects of perception are particular qualities, not species of qualities. So the presentational contents of original conceptions characterise individual properties, not genera or natural kinds to which those qualities belong. Don't confuse original perceptions of secondary qualities, in which a subject receives a presentational content like 'the cause of *that* sensation', with conceptions carrying species-focused presentational contents like 'the kind of thing that causes *that* kind of sensation'. Naive perception alone doesn't even support the claim that the cause of a certain sensation is similar to the cause of another qualitatively similar sensation. It is natural to think that like causes have like effects.[12] Reid acknowledges a 'strong desire to find out connections in things', and so it's tempting to speculate about causal patterns.[13] But such conjectures don't count as perceptions, illusions or hallucinations. They are inductions or educated guesses, products of reason and eagerness to discover natural connections.

In his definitions of secondary qualities, Reid does not mention sensation types or secondary quality types. He merely says, '[Secondary qualities] are conceived only as the unknown causes or occasions of certain sensations with which we are well acquainted'.[14] Moreover, concerning universals, Reid is a nominalist and perhaps a trope theorist,[15] although one need not adopt trope theory to endorse Reid's account of perception. For Reid, you don't perceive heat in general or the Platonic form of heat upon placing your hands into a water bowl. You perceive the heat of *this* water. The conception you develop is of a particular heat of a particular water, and it commits you to nothing about whether this water's heat is by nature similar to that fire's heat. To discover similarities, you must compare the two qualities as they are in themselves, and for that you must know something more about them.

Consider a thought experiment Reid uses to show that conceptions of primary qualities are not formed by considering or combining sensations. Imagine a person with only the sense of touch, whose body is secured and immovable, and who has lost all conceptions of physical qualities. Now give him a sensation, a pain perhaps, and consider his conceptual responses. About him, Reid says,

> Suppose him first to be pricked with a pin; this will, no doubt, give a smart sensation: he feels pain; but what can he infer from it? Nothing surely with regard to the existence or figure of the pin. He can infer nothing from this species of pain, which he may not as well infer from the gout or sciatica. Common sense may lead him to think that this pain has a cause; but whether this cause is body or spirit, extended or unextended, figured or not figured, he cannot possibly, from any principles he is supposed to have, form the least conjecture.[16]

Because the pain communicates little about the nature of its cause, the subject conceives of the pin only as the cause of the pain sensation, just as you originally conceive of secondary qualities only as causes of their respective sensations.

Now subject the experimental subject to another sharp pain, say, a bee sting. He might conjecture that both pains were caused by objects of the same sort, supposing like causes of like effects. But neither the pained subject's sensations nor his naive conceptions of the pin and bee (causes of the pains), force him to think that the two pains have the same cause.

Likewise, an original conception of a rose's smell is just 'the cause of *that* sensation'. If you experience the smell for the first time on Monday and have a similar experience caused by the same rose on Tuesday, you might form two conceptions, one with presentational content 'the cause of the Monday sensation' and the other with presentational content 'the smell of the Tuesday sensation'. Nothing about the smell sensations or these two conceptions lets you deduce that the two causes have a common causal origin in the rose. In fact, they do not even link the sensations to the rose. Your inductive reasoning and your desire to find causal patterns may lead you to conjecture that the two sensations, being very similar, probably have similar causes. But, again, this is not perception.

How might this 'particular quality' interpretation account for the ability to detect correlations among sensations and perceivable qualities? For primary qualities, an easy answer presents itself. Since original conceptions of primary qualities are products of type-3 natural signs, you understand what primary qualities are in themselves. So you notice that the hardness of the desk is the firm adhesion of its parts and that the hardness of the mug is the firm adhesion of its parts, and the desk and the mug have a noticeable similarity because their parts adhere firmly. However, thinking that the hardness of one is like the hardness of the other has little to do with their causing similar sensations. If you accurately conceive of the qualities as they are in themselves, then you think them similar because you understand what they are and can compare their natures.

Secondary qualities are more difficult. Since original perceptions don't reveal the natures of secondary qualities, you cannot immediately tell whether two similar sensations have similar causes or not. If such correlations exist, you have to discover them through everyday experiences, science and other empirical means. Reid describes the process in the rose-smelling case:

> The smell of a rose is a certain affection or felling of the mind; and as it is not constant, but comes and goes, we want to know when and where we may expect it, and are uneasy till we find something, which being present, brings this feeling along with it, and being removed, removes it. This, when found, we call the cause of it.[17]

In this story, you experience a number of similar sensations, so you naturally consider the possibility that they have a common cause. A little experience informs you that the sensations occur when a certain rose is near and cease when the rose is far. So you rightly conclude, on the basis of induction and experiment, that some quality in the rose causes the sensations. But once again, this is not a perceptual belief. Nothing in your original conceptions of the quality tell you that the sensations have a common cause or even similar causes.

Similarities among primary quality particulars become manifest by comparing their natures, not their sensational effects. In fact, a single primary quality can cause multiple sensation types as well, so the sensation-quality correlation fails with primaries just as with secondaries. You can detect a certain desk's hardness by applying pressure to it with, say, your hand. This causes a sensation, which triggers a conception of the desk's hardness. The amount of pressure you apply, and therefore the sensation that you experience, can vary greatly depending on your health, physical strength and other factors. Yet your perceptions do not incline you to believe that the desk grows softer on days when you feel fit and in good health, harder on days when you feel ill or sore. The relationship between sensation and conception is more complex. One quality may cause a variety of sensations. There is no invertible function from one type to the other.

In the case of secondary qualities, it seems even more obvious that one quality may cause a variety of sensations. A bowl of lukewarm water causes warm sensations in some situations, cold sensations in others. A single colour, likewise, seems not to change, although it may cause a variety of sensations. Find a flower in the sunlight, pluck it and bring it into some artificial light. The sensations caused by the flower change a bit, but because of colour constancy, nothing about this suggests that the flower has changed colour.[18] Common sense accepts that a single quality may cause a variety of sensations and that a variety of qualities may cause only one sensation, despite the inclination to ascribe common causes to like effects.

For Reid, speculation and reason, not perception, identify particular qualities as members of natural kinds and propose taxonomies of them. A single physical property can cause of multiple sensations, and, contrariwise, a single sensation can be effected by multiple property types. Correlations between sensation and quality types can be discovered through experience, induction and experimentation. But don't presume that, if sensation-quality patterns are discovered, they will be simple or exactly as hypothesised. Go ahead and find a pattern but don't assume that the first conjecture will succeed.

Jonathan Bennett's PTC

Whereas the water bowl experiment concerns itself with multiple sense experiences in the same subject, Jonathan Bennett considers variations across perceivers. He offers the taste of Phenolthio-urea (PTC) as evidence against the identity theory of secondary qualities.[19] The taste of PTC, he claims, depends not only on the substance's inherent qualities, but also on who perceives it. About 75 per cent of the population finds PTC bitter, but it is tasteless to the other 25 per cent.[20] You can easily imagine possible worlds in which the entirety of the population finds PTC bitter and others in which all think it tasteless. Perhaps humanity could create such worlds through surgery or selective breeding. Bennett believes that this possibility presents grounds for understanding the bitterness of PTC, and tastes as well as secondary qualities generally, as merely mental properties.

Similar to the water bowl example, Bennett traces his argument as follows. If any large group perceives PTC as bitter, then PTC is bitter. If any large group perceives PTC as tasteless, then PTC is tasteless. A large group perceives PTC as bitter. A large group perceives PTC as tasteless. So PTC is both tasteless and bitter. Maybe PTC is bitter to one perceiver and simultaneously tasteless to another, but it cannot be absolutely both bitter and tasteless, unless they are merely mental properties. Thus, the tastes, bitter and tasteless, cannot be physical properties, identifiable with known scientific properties, as Reid suggests.

As in the water bowl case, this argument exhibits confusion between sensation and perception. The premises that different perceiver groups perceive PTC differently are unwarranted. That 75 per cent of the population finds PTC bitter almost certainly means that 75 per cent of the population experiences bitter-type sensations upon touching their tongues to the substance. Some property in the PTC, its taste, causes some sensation in them. They develop a conception of the taste with presentational content 'the cause of *that* sensation', or even 'the cause of *that* bitter-type sensation'. Such a conception makes reference to a bitter sensation, but the sensation is not the same as the conception. This group perceives the PTC's taste as the obscure cause of their bitter-type sensations. They do not perceive the intrinsic natures of the causes of bitter sensations. Again, one quality may cause an array of sensations, and one sensation may have many causes. This group perceives the particular qualities that cause the particular bitter-type sensations they experience.

It will not do to handle the 25-per-cent group like the previous one. To say that this group perceives the quality in the PTC that causes tasteless-type sensations would be silly. There is no such thing as a tasteless sensation. Rather, 25 per cent of the population finds PTC tasteless because they experience no taste sensation upon touching a sample to their tongues. Without a sensation to trigger a conception and belief, how can this group count as perceiving a taste? Apparently, 25 per cent of the population does not perceive the taste of PTC. They are somehow impaired. This suggests a widespread perceptual incapacity, something akin to colour blindness.

How could such a prevalent handicap undermine an identity theory of secondary qualities? Failing to be perceived by large numbers of perceivers does not threaten a thing's existence. If perceiving it isn't terribly important for survival, perhaps evolution can afford to distribute this ability sparingly or not at all. Bennett's treatment of PTC perceptions falls short of undermining the identity theory of secondary qualities as long as secondary qualities bear mere accidental connections to sensations.

Bees and Pigeons

The next case expands on the previous two, addressing dissimilarities in sensations across species. Bees, pigeons and some fish almost certainly have different visual sensations or colour appearances from humans, if they have them at all, since their eye–brain physiology differs so much. Even if scientists found an exact correspondence between sensations and secondary quality types in humans, the correspondence would stop there. Some philosophers have suggested that this cross-species variation makes colour species-relative and therefore subjective and mental rather than physical.[21] However, this move fails for the same reasons as in the previous two cases.

Scientists characterise the human visual apparatus as 'trichromatic', after the three-colour model of perception described in the Young–Helmholtz theory. Human eyes process information from incoming light in three relatively narrow wavelength bands. The high-frequency band is most responsive to 430 nm light waves (blue), the middle band to 530 nm waves (green), and the low to 560 nm (red).[22] Because of this, any colour appearance can be duplicated in a calibrated eye merely by mixing the proper ratio of intensities of these three frequencies of light, as with the Technicolor filtering technique. The brain processes these signals by comparing the stimulation levels of various cone types. Resulting signals represent three features of their stimulation: the overall excitement (how bright things are), the difference between the low and middle frequency cones (red versus green), and the difference between the two lower bands together and the high-frequency cones (yellow versus blue).

Some colour realists wish to correlate sensations with colour properties, but everything in the last few sentences applies only to human beings and a few other species. Other animals, like bony fish, pigeons and finches, have an additional type of cone that responds to ultraviolet light around 370 nm.[23] They have a more discriminating visual apparatus and no doubt a different array of colour appearances from that of humans. Bees use their sensitivity to ultraviolet light to spot bee-appropriate flowers and nectar rich parts of those flowers. A daffodil may look uniformly yellow to you but, if viewing the flower under ultraviolet light reveals anything about the bee's colour appearances, the bee sees bright flowers with dark centres and bright spots at the nectar locations. These spots reflect both ultraviolet and yellow light, which scientists suppose combine to make bee's purple, a colour invisible to humans.[24]

From this information, someone can construct the following argument. If a human perceives the daffodil as yellow, then it is yellow. If a bee perceives the daffodil as bee's purple, then it is bee's purple. A human perceives the daffodil as yellow. A bee perceives the daffodil as bee's purple. It seems that yellow and bee's purple are mutually exclusive, unless they are mental. The daffodil cannot be absolutely both yellow and bee's purple. So the daffodil's colour turns out to be mental and not identical to physical, scientific properties. You might worry that, given the cognitive complexity of perception, bees may not have perceptual abilities. If this is a problem, just substitute some other factive cognitive state, perhaps Dretske's simple seeing, which involves acquiring of information without full blown cognition.[25]

As with the water touching and PTC tasting, responding to the contentions about perceptions of bees and humans is sufficient to derail this objection. You might try to protect the reliability and integrity of human perception, perhaps by claiming that human beings have a corner on the colour market. Sense perceptions of birds and bees mislead them, but not so for humans. This means accepting the human perception of the daffodil's yellow but not the bee's perception of the daffodil's bee's purple. This move, however, is unwarranted. The human visual apparatus is less discriminatory than that of capable birds and bees. If anyone is misled, it is the humans. On the other hand, you could opt to

endorse bird and bee perceptions at the expense of human perceptions as well, but that would undermine the common sense aims
of the present project. You can leave it for the birds, so to speak.

A better approach challenges the notions of yellow and bee's
purple as quality types. They, again, suggest confusion concerning sensation and perception. Although you may have yellow-type
sensations, and these may prompt you to conceive of and believe
in a quality that causes them, the presentational content of your
conception is 'whatever quality caused *that* yellow sensation'. It
does not sort the yellow sensation's cause into some correlative
colour category: red, yellow or blue. Who knows what a bee's
conceptual life is like? But suppose that it forms a perceptual
conception of the daffodil's colour as 'whatever quality caused
that bee's-purple sensation'. Then why not posit a quality in the
daffodil that causes yellow sensations in you and one that causes
bee's purple sensations in the bee? Perhaps these are the same
quality, perhaps not. Regardless, these perceptions offer no reason to think that colour is merely mental. If the daffodil has two
colours, then one is perceivable by humans and the other by bees.
If it has only one colour, then that colour causes yellow appearances in humans and bee's-purple appearances in bees. Either way,
it's wrong to conflate colour types with sensation types, and the
identity theory of secondary qualities remains unthreatened.

Hardin's Viewing Conditions

The previous case may cause you to wonder exactly what colour
daffodils are. Are they yellow? Bee's purple? Both? How do you
name a colour with two colour appearances? Sadly, these are not
easy questions. They are best answered by considering the viewing
conditions case.

C. L. Hardin explains that colours don't reduce to physical
qualities because spectral reflectances, the best candidates for
colour explanantia, do not exhibit a one-to-one correlation
with sensations, even considering 'normal' viewing conditions.[26]
A single object can cause very different colour appearances depending on environmental and background details. There is no good
way to specify which perceptual environment reveals an object's
true colour appearance. Of course, human sense organs require
proper conditions to function well, but Hardin's objection does

not involve unsuitable viewing conditions. None of the situations Hardin considers is suspicious or even unusual.

The first of Hardin's concerns is adaptation. The eye changes its calibration to compensate for light that is especially intense for any particular cone type. Too much red light makes the eye less sensitive to red and more sensitive to green. Greens viewed against red backgrounds appear more intense than against other colours and vice versa. Moreover, normal colour appearances had by a subject looking at a 560nm-reflective object in broad daylight are different from those experienced in dim artificial light, even if that light is incandescent, which, like daylight, has a good saturation across the visible spectrum. If something causes a subject to experience a magenta colour appearance in daylight, a quick transition into artificial light may result in a yellow–red appearance, which fades into unique red as the eye recalibrates.[27]

Think of Hardin as arguing along the lines of previous objections. If you perceive a certain object as magenta, then the object is magenta. If you perceive it as unique red, then it is unique red. You perceive it as magenta and at another time as unique red. But it can't be both magenta and red unless it or its colour is a merely mental property.

It's tempting to say that only one perception, either magenta or unique red, occurs in appropriate viewing conditions. Someone might propose standard background and lighting conditions for all colour evaluations: black background and in bright sunlight for example. This eliminates the commitment to veridical perception in either the magenta or the red case on the grounds that it did not occur under standard conditions. This approach won't work, according to Hardin, because perceptual abilities aren't so limited. They function correctly and usefully in many environments. Human subjects tend to identify an object as the same colour, despite differences in conditions and sensations. A red tomato gets labelled red in daylight, in incandescent light, by candlelight and so on. This is known as 'colour constancy', and it suggests that imposing strict standards for lighting conditions is unnecessary.

Moreover, Hardin claims that colour constancy depends on environmental concerns, like contrasts in a subject's field of vision. The tomato against a green background appears redder than it would against other backgrounds. One might suggest always evaluating colours against perfectly dark backgrounds or backgrounds of the same spectral reflectance as the object, but objects lose their colour constancy.[28] Colour appearances depend not only on lighting conditions but also phenomenological features of adjacent objects. So stipulating standard viewing conditions can't establish a non-arbitrary background against which evaluate a thing's true colour. Hardin makes it clear that an exact one-to-one correlation between sensations and spectral reflections is impossible.

In the case of the water bowl, a single property caused two different sensations at the same time. Other objections emerge because Reid's theory allows a single physical quality to cause different sensation types in different perceivers or different biological species. However, they confuse sensations with perceptions. A similar answer arises for the problem of viewing conditions. A single property can cause multiple sensations yet be veridically perceived according to each sensation. Hardin's argument inappropriately attempts to correlate colour appearances with spectral reflectances when the identity theory accepts that one physical object or spectral reflectance might yield different sensations on different occasions. An object can have a property that causes magenta sensations in some circumstances and unique red sensations in others. So don't conceive of the property as red or magenta, as the objection states, but as causing red and magenta colour appearances. That is, the presentational contents of perceptual conceptions are 'the cause of *that* red-type sensation' and 'the cause of *that* magenta-type sensation', not 'the red colour' or 'the magenta colour'. So you don't really see an object as both unique red and magenta. Rather, first, you see a quality that causes a magenta-type sensation. Then, you see a quality that causes a unique-red-type sensation. It may well be the same quality, for the same quality can cause multiple sensations.

Nothing here is terribly radical. Primary qualities also cause a wide variety of sensations. Reid considers the case of viewing a book:

> Every one who is acquainted with the rules of perspective, knows that the appearance of the figure of the book must vary in every different position: yet if you ask a man that has no notion of perspective, whether the figure of it does not appear to his eye to be the same in all its different positions? he can with a good conscience affirm, that it does. He hath learned to make allowance for the variety of visible figure arising from the difference of position, and to draw the proper conclusion from it.[29]

Not only does the book's shape cause a variety of sensations, but the typical perceiver interprets the sensations without difficulty. Each sensation or appearance indicates book's one and only, categorical shape. Primary qualities cause multiple sensations, and this poses no barrier for accepting them as objective physical properties. Surely then, the fact that each secondary quality can cause a variety of sensations shouldn't count as evidence for non-physicality. Just as a physical taste and a physical temperature can cause many sensations, so can a physical colour.

Some will find this a curious response, since it offers no perceptual method for determining the actual colour of a perceived object. How can anyone figure out what colour the object is, magenta or red? Is the daffodil yellow or bee's purple? As G. J. Warnock says, 'Consider a simple case of . . . seeing a quality, for example seeing the colour of Lloyd George's tie. It is clear that one could not rightly say that one saw the colour of his tie, if one did not get to know at the same time what colour it was.'[30] Doesn't perceiving a thing reveal what colour it is? Well, not exactly.

Again, return to Reid's account of the presentational content available in a sensory conception of a secondary quality. In the case of the red-type appearance, your senses lead you to conceive of some quality, which you conceptualise as the cause of the appearance. In the magenta case, you consider the cause of the magenta-type appearance, which happens to be the same quality as in the red case. What you do not find in these presentational contents is any claim about the natural kinds or colour types that these qualities instantiate. You should not expect to find one, so long as you are confined to the deliverances of sense perception. That is why Reid wants scientists to investigate secondary qualities. Perceptual notions of them are obscure. Reid should take issue with Warnock's claim.

Reid, who writes without the advantages of nineteenth- and twentieth-century colour science, suggests that an approach like Warnock's confuses colours with colour appearances, qualities with sensations. About colour types, Reid says,

> When we think or speak of any particular colour, however simple the notion may seem to be, which is presented to the imagination, it is really in some sort compounded. It involves an unknown cause, and a known effect. The name of colour belongs indeed to the cause only, and not to the effect. But as the cause is unknown, we can

form no distinct conception of it, but by its relation to the known effect. And therefore both go together in the imagination, and are so closely united, that they are mistaken for one simple object of thought.[31]

Reid thinks that since colour perceptions offer no information about their natures, some like to inflate the association between colour and sensation. Perceivers do not understand the natures of colours by perception alone. Without scientific inquiry there are no grounds on which to organise colours, no way to build a taxonomy, except by the sensations that they cause. Reid continues by noting examples,

> When I would conceive those colours of bodies which we call scarlet and blue; if I conceived them only as unknown qualities, I could perceive no distinction between the one and the other. I must therefore, for the sake of distinction, join to each of them in my imagination some effect or some relation that is peculiar. And the most obvious distinction is, the appearance which one and the other makes to the eye.[32]

If it were not for variations in sensations, there would be no way whatsoever to sort colour qualities, unless someone learns more about their natures. Names for supposed colour types, like 'blue' or 'scarlet', don't characterise the qualities as they are in themselves. And this fact makes Western traditional colour categories somewhat dubious. Perhaps they do not accurately reflect variations in the natures of colours. As Reid puts it,

> [T]he appearance is, in the imagination, so closely united with the quality called a scarlet-colour, that they are apt to be mistaken for one and the same thing, although they are in reality so different and so unlike, that one is an idea in the mind, the other is a quality of body.[33]

You may conjecture that the natures of colours correlate with your colour sensations. But in reality, colour sensations are not colours, nor are they similar to colours. So keep in mind that your conjecture might be false, a point raised by Descartes as well.[34]

How can it be that you see the colour of Lloyd George's tie without knowing what colour the tie is? Colour distinctions are not given to you in original perceptions of colours, apart from distinctions among the appearances that the colours effect. While seeing the colour of the tie allows you to say what sensation the tie causes, your attempts to specify the tie's colour as green or blue can fail. Green and blue belong to a hypothetical colour taxonomy based on sensations that colours cause, not the objective natures of the colours. You perceive the colour of a certain object as that which causes a magenta sensation at one time and that which causes a unique red sensation at another. There is no reason to think that the same quality could not cause both. The viewing conditions objection mistakes appearance types for colour types.

Hardin's Metamerism

Like the previous two cases, the final non-correspondence objection involves colours. For colours, not only is there no function from scientific property to sensation, but there is also no function from sensation type to scientific property. Two objects can exhibit the same colour appearance in the same conditions, even though they have very different physical properties. Whereas the last section considered colours that cause multiple appearances, here the worry arises from a single colour appearance type caused by a variety of physical properties. This phenomenon, known as 'metamerism', emerges as an inevitable consequence of human trichromatic vision. And, again, C. L. Hardin finds it a reason to reject colours as physical properties.[35]

Again, the problem arises because the cones in human eyes come in only three types, each of which responds to a relatively narrow band (a domain of frequencies) of light. The information carried along the optic nerve, and hence to the brain–mind, consists of only two colour signals: one that represents the difference in the stimulation of the green and the red cones and another that communicates the difference between the blue cones and the sum of the others, which together account for yellow. There is a third signal that communicates about light and dark, but ignore it for the moment.[36]

It's easiest to understand metamerism by trying an example. Suppose that you're under typical daytime lighting conditions looking at an object that reflects only 540 nm light. It has a bright-yellow appearance since your red and green cones are about equally stimulated and the blues hardly at all. Now consider a second object that reflects equal amounts of 530 and 560 nm light. Again your green and red cones are equally stimulated, and the blue cones are left out. Assuming that you have about the same concentration of green and red cones, this object also appears bright yellow. In bright daylight and other white lights, the two objects exhibit indistinguishable colour appearances. They have different scientific properties, different spectral reflectances and micro-physical bases of those reflectances, but cause the same type of sensation. Similar scenarios can be generated for practically any colour appearance. Thus, it seems that a thing's colour can't be identical to its spectral reflectance. Colour appearances do not correspond to spectral reflectances.

The metamerism objection to secondary qualities as physical properties goes like this. First, it says that what is true of colours generally is also true for each species of colour. Take yellow as an example. If colours are physical properties, then yellow is a physical property. This is necessarily true if yellow is a colour species. Now consider two objects, which both appear yellow. If yellow is a physical property, then the objects share microphysical surface structures or spectral reflectances. But what is the structure or reflectance? Experts have looked for it and found nothing but tremendous complexity.[37] In fact, the objects lack common features in their surface structures and spectral reflectances. So the yellowness of these objects can't be a physical property, and furthermore neither can colour. Once again, if colour is not scientific, then the identity theory fails.

In answering the metamerism objection, it's tempting to target the premise that the objects do not share a scientific property. For example, maybe scientists haven't yet compared the spectral reflectances in terms of a single average or some other coarse-grained or even disjunctive category. But these efforts are to no avail. Averages do not account for the way that cone cells and the optic nerve transmit information to the brain. Even in the above example, the average wavelength reflected by the second object $((530 \text{ nm} + 560 \text{ nm})/2 = 545 \text{ nm})$ is not the same as that reflected by the first (540 nm). If one object reflected 540 nm light and another reflected both 540 nm and 10 nm light (average 225 nm), the two would look the same, since you have no biological mechanism for detecting X-rays. Coarse-grained or disjunctive colour properties, once adapted to fit the data, are so convoluted that it's hard to

count them as natural kinds. There are potentially infinite dissimilar wavelength combinations that can cause yellow appearances. By way of analogy, J. J. C. Smart compares course-grained
and disjunctive understandings of colour species to the property
of snarkhood, which, he stipulates, is being a tomato or rainbow
or bulldozer or archbishop.[38] The coarse-grained and disjunctive
approaches make colour kinds too ad hoc, too contrived.

Michael Huemer hopes that one day scientists may discover the
physical property that links the two yellow objects.[39] Right now,
scientists lack good theories of what it means to be red, blue or
yellow, but this, he says, does not mean that they cannot eventually find one. The problem with Huemer's approach is that it is not
needed. The theory of trichromatic vision already says why human
beings experience the same colour appearances in response to different physical properties. It is an expected consequence of ocular
physiology. Huemer seems to think that scientists will find some
non-physiological commonality among metamers, some additional explanation for why different qualities cause similar sensations, but this multiplies metaphysical properties beyond necessity.

Fred Dretske tries to label one object as the really-yellow object.
He hypothesises that metamerism is the result of employing colour
vision outside of its proper context.[40] This means that things can
look yellow without being yellow. Yellow, Dretske thinks, is whatever property that yellow-perceiving ability is 'designed' to detect,
that property the detection of which originally gave early humans
or proto-humans the relevant evolutionary advantage.[41] In this
case, if an object has the property that the human yellow-perceiving
ability is meant to detect, then it is yellow and other types of
objects aren't.

There are at least two undesirable features in Dretske's
proposal. The first is that it leaves too many questions about non-
yellow objects that yellow colour appearances and seem to be
coloured. According to Dretske, they aren't yellow. So are they
another colour? If so, which one? Or are they somehow colour-
less, or not perceivable? Such questions introduce drawbacks to
Dretske's theory but perhaps not insurmountable ones. A worse
problem is the theory's proliferation of colour illusions, since it
makes colour perception into a detector for finding some formerly
evolutionarily advantageous property. If the yellow-sensing ability
arose in order to detect bananas, then your yellow appearances are

supposed to alert you to the presence of bananas. But if this is the case, then your banana detector seems ridiculously oversensitive. It responds to sunflowers, pedestrian crossing signs and most rubber duckies. According to Dretske, each of these responses is an illusion, because yellow colour appearances only suit bananas. Every time you see an old-fashioned rain slicker, your senses try to dupe you into thinking that it is a banana. Direct Realism can tolerate some illusions, perhaps even as many as Dretske effectively suggests. But better not have them.

The Reidian theory of perception offers another possibility: denying yellow as a colour species. The objection says that, if colours are physical properties, then yellow is a physical property. True if yellow is a species of colour. But is yellow a species of colour? Consider how Reid might approach the problem.

According to Reid, common sense tells perceivers that colours are physical, but perceptions of objects' colours do not reveal whether they are yellow or even the same colour.[42] To say otherwise confuses sensations with perceptions. By visual perception alone, the presentational content is 'whatever quality caused that yellow sensation'. Nothing here entails that two yellow appearances belong to the same yellow-sensation-causing physical property. One might hypothesise that the colour quality in one object is like the colour quality in the other, since like effects often indicate like causes. But this is conjecture, not perception. If it is erroneous, then it is the fault of intellectual hastiness, social conventions or inductive reason, not perception. As far as perception is concerned, yellow need not be a species of physical property. It can be a species of colour appearance, caused by more than one physical property.

On Reid's theory of perception, it is quite possible for a natural sign or sensation to be caused by multiple properties. This is even true for primary qualities, especially when perceived by acquired perceptions. Consider again Reid's example of viewing a book's size and shape. The book's visible figure and magnitude serve as signs (although they are not sensations) for its intrinsic figure and magnitude.[43] That is, the two-dimensional projection of the book's silhouette on your eye leads your mind to consider the book's intrinsic dimensions.[44] About perceptions of the book's magnitude, Reid says,

> Whether I view [the book] at the distance of one foot or of ten feet, it seems to be about seven inches long, five broad, and

one thick. I can judge of these dimensions very nearly by the
eye, and I judge them to be the same at both distances. But yet
it is certain, that at the distance of one foot, its visible length
and breadth is about ten times as great as at the distance of ten
feet; and consequently its surface is about a hundred times as
great.[45]

Many visible figures and magnitudes can represent the book's one
intrinsic size and shape. Reid's claim about the visible dimensions
of the book from a distance of one foot suggests that the reverse
may also be the case. Many intrinsic sizes and shapes may be represented by a singular type of visible figure. At an arm's length, the
most recent edition of Reid's *Inquiry* has the same visible dimensions as Robert Swartz's *Perceiving, Sensing, and Knowing* when
viewed from ten inches nearer. And Swartz's book, when viewed
at arm's length, has a visible height roughly equivalent to a nine-
storey hospital about a block away. Thus, a single visible height
serves as a natural sign for real heights of both seven inches and
nine storeys. Likewise, a single sensation may stand for multiple
qualities.

For those who are not squeamish about the compatibility of
scientific realism and nominalism about universals, another way to
cast this solution is by way of Reid's trope theory. Colour qualities
are objects of perception, causes of sensations. But to be causal
and to be particular objects of perception, they must be tropes, not
generalisations, genera or categories. As Reid puts it, 'The whiteness of this sheet is one thing, whiteness is another'. So members
of a species differ from the species itself. He goes on, '[The whiteness of this sheet] signifies an individual quality really existing . . .
[Whiteness] signifies a general conception which implies no existence'.[46] For Reid, qualities are particular features of particular
entities, not types into which those entities fall. It is these particular qualities that exist, are causally potent, and serve as objects of
perceptions.

Since, for Reid, the objects of perception are causal, they must
be particulars. Even if two objects are yellow, it is not yellowness in general that perceivers see. It is the particular colour of an
object, which is a feature of that object and not the other, and the
particular colour of the second object, which is a quality of the
second and not the first. Particular qualities are the true objects of
perception.

The colour of a perceived object is whatever property accounts for the particular colour appearance you have when you look at it. Almost certainly that means, as Reid puts it, 'the disposition of bodies to reflect a particular kind of light'.[47] The colour of one object might be its disposition to reflect 540 nm light, the colour of the second is its disposition to reflect 530 and 560 nm light at the same time. Each disposition is constantly conjoined to a yellow appearance, and so causes them in the 'vague and popular' Humean sense.[48]

'Yellow' is an attempt to generalise property types based on sensations. It is not a property of any particular entity. It is not possessed by any object, and Reid even says that such generalisations do not exist.[49] Without existing and being possessed by particular bodies, there's no way that yellowness in general can be involved in causation. And since objects of perception become objects of perception by causing sensations, yellow in general can't be an object of perception. Yellow isn't what you see when looking at the two objects. You see the colours of the two objects, which both happen to cause yellow appearances. Colours cause physiological events and belong to physical objects. Yellow in general does not, and therefore it is not a colour. The colours of the two objects are their respective spectral reflectances. Yellow, on the other hand, is a failed conjecture about those colours perhaps belonging to a hypothetical natural kind or colour species.[50]

The drawback to the Reidian solution is that it makes yellow neither a colour nor a natural kind of colour. The same goes for red, blue, green and others in between. The standard Western colour taxonomy consists of mistakes, erroneous speculations and hasty generalisations. Similar sensations do not imply similar physical kinds. You might notice that a daffodil occasions a sensation similar to that caused by a rain slicker, but inferring that the two are the same colour is going too far. Supposing them similar is not unreasonable, but it is epistemically hazardous, like Cartesian ill-considered judgements.[51]

This conclusion is not as surprising as it may seem. Inaccurate generalisations are quite common in human thought. Take an example in the field of biology. The now-extinct Tasmanian wolf (*Thylacinus cynocephalus*; Fig. 7.1) looks much like the grey wolf (*Canis lupis*; Fig. 7.2).

Figure 7.1 Tasmanian wolf

Figure 7.2 Grey wolf

When classifying animals by natural kinds, it's tempting to put these two animals in the same family. Each has large pointed ears at the top of the head, binocular vision, a long narrow face, strong shoulders, hock in the hind leg, long tail and coarse fur. The most prominent differences, size and fur pattern, seem insignificant for purposes of phenotypic classification.

Despite appearances, these species are not closely related. The Tasmanian wolf is a marsupial, whereas the grey wolf is placental. Genetically, the Tasmanian wolf has more in common with koalas than canines, and the grey wolf shares a common ancestor with weasels and house cats before it has any links to its Australian lookalike. Their common physical features are explained by parallel evolution, not genetic relations. The inclination to lump these 'wolves' together is an error in induction, an ill-considered judgment.[52]

As a few other philosophers – for example, Keith Allen and Mazviita Chirimuuta – claim, Western names for supposed colour species in fact fail to identify natural kinds of colours.[53] Some will find this a high price to pay for a solution to the Problem of Secondary Qualities. 'Yellow must be a colour if anything is', they will think. But there is good reason to doubt this sentiment. Psychologists comparing the colour vocabularies of English, Berinmo (of Papua, New Guinea), and Himba (of Southern Africa) have found that colour vocabulary and taxonomies differ greatly among cultures.[54] Like race, human colour categories are culturally and linguistically defined, not inherent in physical objects or human perceptions. Moreover, dogmatic adherence to the traditional Western colour taxonomy is ethnocentric, treating non-Westerners as if they have a perceptual impairment. There are good scientific and moral reasons to keep a loose grip on the commitment to yellow as a colour.

In light of contemporary colour science, Reid's theory of secondary qualities leads to a curious combination of claims regarding the natures of colours. First, colours are physical properties. Traditional categories fail because, second, colours are referential contents or objects of perception even though their natures are not correctly or carefully understood. You can see a thing's colour, which is probably identical to its spectral reflectance, but your visual organs do not take in enough information to determine, merely by looking, what colour the thing is. Third, as it turns out, traditional colour names – 'red', 'yellow', 'green' and 'blue' – fail to pick out natural kinds. These mere conventions do not necessarily reflect anything about the objective natures of colours.

There is no invertible function from subjective sensation to objective physical property or vice versa, no exact correlation between quality and appearance. However, this is no reason to exclude secondary qualities, the causes of secondary quality sensations, from the catalogue of scientific properties. It merely means that perception's understanding of them is obscure and relative.

Reid's theory clearly distinguishes sensation from perception. The vague presentational contents of secondary quality perceptual conceptions diffuse the threat of non-correlation. Instead, the causal relationships between sensations and physical properties become objects of scientific wonder.

Notes

1. Locke, *An Essay concerning Human Understanding*, 2.8.14/137, 4.3.11/544.
2. Locke, *An Essay concerning Human Understanding*, 2.8.10/135, 2.8.14/137, 2.8.24/141, 2.8.22/140.
3. Descartes, *Principles of Philosophy*, 1.70/218.
4. Locke, *An Essay concerning Human Understanding*, 2.8/139.
5. Huemer, *Scepticism and the Veil of Ignorance*, pp. 140–5.
6. Tye, *Consciousness, Color, and Content*, pp. 150–65.
7. Dretske, *Naturalizing the Mind*, p. 93.
8. Tye, *Consciousness, Color, and Content*, pp. 152–3.
9. Locke, *An Essay concerning Human Understanding*, 2.8.21/139.
10. George Berkeley, *Three Dialogues*, pp. 14–15.
11. Reid, *Inquiry into the Human Mind*, 2.9/42. There are passages in which Reid seems to endorse a sensation-quality correlation (*Inquiry into the Human Mind*, 4.1/49). If this is the case, then perhaps Reid is inconsistent on this point.
12. Hume, *An Enquiry Concerning Human Understanding* 7.2/49–53.
13. Reid, *Inquiry into the Human Mind*, 5.9/41.
14. Reid, *Essays on the Intellectual Powers of Man*, 2.17/202.
15. Reid, *Essays on the Intellectual Powers of Man*, 4.2/323; Reid, *Essays on the Intellectual Powers of Man*, 5.1/356, 5.3/367. Reid is cast as a trope theorist in Laurence and Margolis, 'Abstraction and the Origin of General Ideas'; Lehrer and Tolliver, 'Tropes and Truth'.
16. Reid, *Inquiry into the Human Mind*, 5.6/65.
17. Reid, *Inquiry into the Human Mind*, 2.9/40.
18. Tye, *Consciousness, Color, and Content*, pp. 147–8.
19. Bennett, 'Substance, Reality, and Primary Qualities'.
20. For a historical perspective on this phenomenon, see Wooding, 'Phenylthiocarbamide'.
21. See Allen, 'Inter-Species Variation in Colour Perception' for a good summary of this position, although the paper ends up endorsing a view closer to Reid's.
22. Berns et al., *Billmeyer and Saltzman's Principles of Color Technology*; Bear, Connors and Paridiso, *Neuroscience: Exploring the Brain*, ch. 9.
23. Hart et al., 'Visual Pigments, Cone Oil Droplets and Ocular Media'.

24. Raven, Evert and Eichhorn's *Biology of Plants*, pp. 532–3. This text contains a helpful photograph of some angiosperms under ultraviolet light, highlighting the contrasts a bee must notice as it approaches the flower – light petals, dark centre and bright 'honey guides' near the nectar. By contrast, to human perceivers the same flower looks uniformly yellow under regular sunlight.

25. Dretske, *Perception, Knowledge and Belief*, pp. 97–112.

26. Hardin, 'A Spectral Reflectance Doth Not a Color Make', pp. 191–202.

27. Hardin, 'A Spectral Reflectance Doth Not a Color Make', p. 193.

28. Hardin, 'A Spectral Reflectance Doth Not a Color Make', p. 195.

29. Reid, *Inquiry into the Human Mind*, 6.3/83–4.

30. Warnock, 'Seeing', pp. 59–60.

31. Reid, *Inquiry into the Human Mind*, 6.5/86.

32. Reid, *Inquiry into the Human Mind*, 6.5/86–7.

33. Reid, *Inquiry into the Human Mind*, 6.5/87.

34. Descartes, *Meditations*, 7.81/56–7.

35. Hardin, *Color for Philosophers: Unweaving the Rainbow*, pp. 45–51, 67–75.

36. Berns et al., *Billmeyer and Saltzman's Principles of Color Technology*.

37. Clark, *A Theory of Sentience*, p. 213.

38. Smart, 'On Some Criticisms of a Physicalist Theory of Colors'.

39. Huemer, *Scepticism and the Veil of Ignorance*, pp. 140–5.

40. Dretske, *Naturalizing the Mind*, pp. 88–93.

41 Dretske credits D. R. Hilbert with the same view. See Hilbert, 'What is Color Vision?', p. 362.

42. Reid, *Inquiry into the Human Mind*, 6.5/86–7.

43. Reid, *Inquiry into the Human Mind*, 6.7/97.

44. Reid, *Inquiry into the Human Mind*, 6.2/79–82, 6.7/95–8, 6.9/103–12.

45. Reid, *Inquiry into the Human Mind*, 6.4/84.

46. Reid, *Inquiry into the Human Mind*, 5.3/367.

47. Reid, *Essays on the Intellectual Powers of Man*, 2.17/204.

48. Reid, *Inquiry into the Human Mind*, 2.9/40.

49. Reid, *Inquiry into the Human Mind*, 2.9/40.

50. Shrock, 'Yellow is Not a Color'.

51. Descartes, *Meditations*, 7.82/56.

52. Paddle, *The Last Tasmanian Tiger*, pp. 2–8.

53. Keith Allen, 'Revelation and the Nature of Colour'; Chirimuuta, *Outside Color*, p. 185.

54. Roberson et al., 'Color Categories'.

8

A Priori Objections

Since Reid's day, two challenges to the identity theory of second-ary qualities object on purely philosophical grounds. The first, promoted by Howard Robinson, criticises the causal connection between physical and mental, external property and subjective phenomenon. Second, Bertrand Russell complains that differing knowledge conditions for secondary qualities and physical prop-erties prevent the identity theory from meriting serious consider-ation. Consider them in that order.

Robinson's A Priori Sufficiency

As in the preceding case, the next two objections confuse sensa-tions with perceivable properties. Part II offered an account of secondary qualities whereupon they count as physical without becoming causally superfluous. How? Secondary qualities are identifiable with certain physical, scientific properties. Howard Robinson, who featured in Part I, says that such a proposal is impossible because no candidate for identification is a priori suf-ficient to explain 'phenomenal' secondary qualities. Perhaps his criticism communicates best in light of some observations about scientific explanations in general.

Consider a paradigm case of scientific identification. The phases of matter – solid, liquid and gas – may be specified as microphysi-cal properties relating to chemical bonding. Firm bonds create the rigid structures of solids, whereas weak or non-existent bonds cause substances to behave as liquids or gases. One reason that such an explanation works is that its truth entails the phases of matter. That is, the explanans necessitates the explanandum. It is impossible for the molecules of a substance to bind rigidly without the substance becoming a solid, and likewise for other bonding strengths and phases. As Robinson puts it, the explanans of chemical bonding strength is '*a priori* sufficient' for the explanantia of the material phases. He explains,

> If the molecules bind in a certain way the object just cannot, for example, behave as a liquid. This is not an empirical truth, for if the molecules are binding tightly, that means that they don't move easily relative to each other, which entails that the object is not flowing.[1]

A solid has certain dispositional properties: it does not flow, and it has a certain elasticity, plasticity, tensile strength, shear strength and so on. An object held together by strong electrochemical bonds always exhibits these mechanical properties. That is, given that an object is composed of molecules with strong electrochemical bonds and arranged in a certain pattern, it cannot help but exhibit the same macrophysical dispositions and other properties characteristic of solids.

Robinson's point is epistemic. In considering possible candidates for theories to explain solidity, scientists might consider the presence of some Greek element like earth or very tiny fairies that draw the parts of the solid object together. If the fairies were strong enough, then their pull would be a priori sufficient for at least some of the solid object's mechanical powers. The coolness and dryness of the earth would explain the solid object's lack of flow. Only after extensive experimentation or taking additional considerations, like simplicity, into account, do they favour the electrochemical theory over these.

To bolster Robinson's case, consider proposals for explanations that fail to meet the a priori sufficiency requirement. Suppose you look for an explanation for the solidity of some object and someone suggests that electrochemical repulsion is present among the object's molecules, or that the Greek element of water permeates the object, or that a giant squid is hiding behind a Volkswagen in Wisconsin. You would reject these out of hand, because of their insufficiency to account for the object's powers and other properties. Electrochemical repulsion among an object's microphysical parts, might explain an object's falling apart or sublimating, but not its tensile strength. Water would explain flowing, but not elasticity. And the squid seems to have no bearing whatever on solidity. Isn't a priori sufficiency a requirement for any a posteriori identity of the sort Reid proposes?

According to Robinson, secondary qualities fail this requirement. While electrostatics can bind atoms into solid objects, there is not the same a priori connection between microstructure or spectral reflectance or effluvia and 'phenomenal secondary qualities'. The explanandum is not entailed by any explanans scientists might propose. Why should fast-moving molecules cause heat perceptions? Why should 440 Hz yield aural perceptions of concert A? Why should 540 nm light cause yellow appearances? When it comes to secondary qualities, identity doesn't seem to hold up.

Robinson's criticisms amount to an objection along the following lines. For an identity of secondary qualities with physical properties to succeed, physical properties would have to be a priori sufficient to account for the manifest features those qualities. Being a priori sufficient for the manifest features of secondary qualities includes entailing the phenomenal aspects of those qualities. However, physical properties aren't a priori sufficient, so the identity theory fails.

This argument is valid. Although controversial among philosophers,[2] it seems right that explanans should entail explanandum in an identification. In fact, this rule resembles a requirement that Reid imposes on explanations. Following Isaac Newton, he says, 'When men pretend to account for any of the operations of nature, the causes assigned by them ought . . . to be sufficient to produce the effect'.[3] In fact, Reid uses this principle in a manner similar to Robinson's argument to dismiss the materialist theory of the mind:

> matter and motion, however subtilly divided and reasoned upon, yield nothing more than matter and motion still.
>
> It would, therefore, be unreasonable to require that [Dr. Hartley's] theory of vibrations should, in the proper sense, account for our sensations. It would, indeed, be ridiculous in any man to pretend that thought of any kind must necessarily result from motion, or that vibrations in the nerves must necessarily produce thought, any more than the vibrations of a pendulum.[4]

Nothing about physical vibrations entails anything about the operations of minds, so Reid thinks it would be 'ridiculous' to posit a theory that purports to explain the nature of thought by physical movements. Out of charity, then, Reid ascribes to David

Hartley the view that physical movements relate causally to certain operations of the mind. But even this theory, Reid thinks, fails the sufficiency test. Compare the explanatory power of vibrations applied to the wide variety of human sensations:

> I know two qualities of vibrations in a uniform elastic medium, and I know no more. They may be quick or slow in various degrees, and they may be strong or weak in various degrees; but I cannot find any division of our sensations that will make them tally with those divisions of vibrations.[5]

Sensations, Reid says, vary by degree as well as by mode. Vibrations, on the other hand, vary in only two respects. So, unless vibrations can exhibit at least as much variety as sensations, Reid does not see how vibrations alone could causally explain them. If Hartley wants to identify all sensations with vibrations, then Reid rules out Hartley's proposal without subjecting it to scientific inquiry, because of its explanatory poverty. Vibrations of the nervous system are, by themselves, not candidates for explaining sensations as physical.

There is good reason to think that Reid concurs with Robinson that physical properties don't entail sensations. Concerning the rose-smelling case, Reid discusses the train of thought from secondary quality sensation to cause, saying, '[The sensation] has no similitude to any thing else, so as to admit of a comparison; and therefore he can conclude nothing from it, unless perhaps that there must be some unknown cause of it'.[6] Nothing about small particles entering a nose entails olfactory sensations. Nothing about 440 Hz vibrations entails concert-A aural sensations. Robinson seems to be correct in saying that there are no a priori connections between subjective secondary quality phenomena and the objective, physical properties.

One phrase, however, draws suspicion, 'phenomenological aspects of secondary qualities'. As Reid insists, perception characterises secondary qualities as physical properties, not sensations. Nor are perceptions of secondary qualities the same as secondary quality sensations. But Robinson's characterisation, 'phenomenal secondary qualities', essentially ties secondary qualities to sensations. He means that an explanation of secondary qualities should explain their association with certain sensations. If so, then why not reject this requirement? After all, on Reid's view the causal connection is a contingent one and more closely linked to the nature of the human mind than to the secondary qualities themselves.

On the other hand, Reid's identification programme fulfils the a priori requirement in other respects, because the conceptions of secondary qualities obtained through sense perceptions aren't the only conceptions available. Recall, for example, the progression of conceptions of the smell of a rose. The presentational content of the conception had by unlearned sense perception characterises the smell merely as the unknown cause of a certain sensation. A more experienced, but still-common-sense conception posits a causal chain from the smelly object to the air to one's nose.[7] A third, more theoretical conception of the smell relates it to tiny airborne particles released from the object and taken in through the nose. Eventually, Reid hopes someone will discover the objective nature of the smell by developing another, more scientific conception.

Notice for a moment the third conception, the one that considers the smell as tiny particles inhaled through the nose. There appears to be an a priori connection between it and the second conception, which merely links the smelly object and the nose. If a theoretician proposed an alternative third conception on which particles issue from somewhere besides the smelly object, say, from one's own head or the moon, then it would offend the second conception, which explicitly locates the smell in the smelly object. Likewise, in light of the second conception, no third conception could specify very large, wider-than-a-nostril-sized carriers of the smell. How would they enter the nose? On the other hand, tiny particles inhaled via the nostrils do a priori necessitate something about a causal chain from smelly object to air to nose. So there are a priori connections between the third, speculative conception and the second. Even when it comes to secondary qualities, some hypotheses may be ruled out on a priori grounds.

Notice the same sort of connection when considering the nature of heat. Sense alone suggests that there is some quality causing a particular heat sensation. Again, on Reid's theory, there is nothing that necessitates the quality's causing that particular sensation. But in fact, it does cause it. Knowing that heat is rapid molecular motion does not reveal why heat causes the sensations it does. But experience can give perceivers additional conceptions of heat as possessed by some physical object; transferable via radiation, conduction or convection; and capable of initiating chemical reactions and so forth. Whatever the attempt to account for the nature of heat, then, the proposed explanation must possess an a priori connection to these common sense characteristics. The caloric, phlogiston and molecular motion theories of heat all offer such connections. Caloric, for example, is supposed to be self-repellent, so it will flow from warmer bodies into cooler ones. And this feature is a priori sufficient for explaining heat conduction. That is, if heat is an element with a disposition to self-repel, then it should behave in the same manner as heat, spreading from warmer objects to cooler. Molecular motion, likewise, is supposed to spread as excited molecules impart kinetic energy to adjacent molecules. But if someone were to suggest that heat is effluvia entering one's nose or pink leprechauns guarding their gold, scientists would dismiss these proposals because they do not account for the conductive features of heat.

Robinson is right about there being no a priori connection from theoretical conceptions of secondary qualities to sensations. Reid's theory allows the consideration of secondary qualities apart from the sensations they cause. However, for Reid, relations between sensations and causal properties are merely accidental. Additionally, sensory conceptions of secondary qualities carry very little information. Once further conceptions develop, through experience and scientific investigation, perceivers discover that there are a priori connections between explanantia and explananda of the sort that Robinson demands. Reid honours the a priori sufficiency requirement in scientific identifications and renders the requirement moot for original perceptions.

Russell's Knowledge by Acquaintance

Some object to the identity theory on the basis of so-called perceptual knowledge of secondary qualities. These objections also seem to exhibit confusion between secondary qualities and the sensations that they cause. For example, Bertrand Russell claims that true knowledge of secondary qualities is essentially first-hand, which he terms knowledge 'by acquaintance'. But knowledge of physical properties is 'by description' rather than 'by acquaintance'. Thus, he finds secondary qualities to be mental and subjective.

Considering light, which isn't among Reid's secondary qualities, Russell explains,

It is sometimes said that 'light is a form of wave-motion', but this is misleading, for the light which we immediately see, which we know directly by means of our senses, is not a form of wave-motion, but something quite different – something which we all know if we are not blind, though we cannot describe it so as to convey our knowledge to a man who is blind. A wave-motion, on the contrary, could quite well be described to a blind man, since he can acquire a knowledge of space by the sense of touch; and he can experience a wave-motion by a sea voyage almost as well as we can. But this, which a blind man can understand, is not what we mean by light, we mean by light just that which a blind man can never understand, and which we can never describe to him.

When it is said that light is waves, what is really meant is that waves are the physical cause of our sensations of light.[8]

Russell drives a conceptual wedge between secondary qualities and physical properties on the basis of human understanding. He thinks that a blind person cannot understand what 'light' means, since he has never experienced light for himself. Hearing a scientist's story about wave motions of photons, Russell thinks, only gives the blind man an understanding of light's causes, not its nature. If the blind man suddenly received sight, he would learn something about light that he did not know before, namely, what light is in itself.

Russell effectively makes the following argument. First, he claims, by taking a sea voyage, a blind person can understand waves, indicating that, for the perceptually impaired, waves are comprehensible. However, the seagoing blind person cannot see light, and so cannot acquire perceptual knowledge of it. His ability to understand light, then, is impaired along with his sight. From here, Leibniz's Law dictates that waves cannot be identical to light, since they have a property that light lacks, namely, comprehensibility by the blind-from-birth.

The argument is valid, and you can easily construct parallel arguments for Reid's secondary qualities. Russell is obviously right about the blind person's ability to comprehend waves, since there are perceptually impaired scientists and mathematicians. The incomprehensibility of light to the lifelong blind receives support from several highly regarded philosophers. In addition to Russell, the seventeenth-century philosopher Pierre Gassendi holds that secondary qualities depend not only on combinations and interactions of atoms and their qualities, but on those atoms cooperating with one's senses.[9] Likewise, C. D. Broad claims that secondary qualities' appearances are essential to their perceiver-independent natures.[10] And John Foster writes that secondary qualities essentially feature in the content of experience in a way unlike primary qualities.[11] Recall from Chapter 4 that some have attempted to read Reid as subscribing to secondary quality knowledge by acquaintance. These writers see secondary qualities as essentially tied to sensations.

By now, however, it's all too obvious how Reid might object to Russell. Sensation and perception are distinct, and the presentational contents of secondary quality perceptions do not say anything about the natures of secondary qualities. The knowledge-by-acquaintance objection claims that normal perceivers understand a secondary quality's nature in virtue of having perceived it. On Reid's account, that is not so, for, although the perceiver has a conception of the quality, the presentational content of that conception is 'the cause of this sensation'. It does not inform the perceiver of the quality's nature. The normal perceiver and the lifelong blind are both, at least initially, ignorant of the quality's nature. The blind subject, failing to have the relevant sensations, may never have a conception by means of perception. But the normal perceiver might inform him of her perception, allowing the impaired subject to form a non-perceptual conception of the quality, perhaps one with presentational content, 'whatever *she* perceived'. This is a weak understanding indeed, but an understanding nonetheless. Through scientific investigation, the normal perceiver and the impaired person may come to a new conception of the quality, one that permits both of them to understand its nature better. But neither has an advantage over the other in conceptualising the quality scientifically, because the scientific conception has nothing to do with the subject's sensations.

In Reid's work, there appears a similar case in his analysis of visible figure. Visible figure, Reid thinks, may be perfectly understood by a blind person and the blind mathematician Dr Saunderson in particular. Reid writes,

> We may venture to affirm, that a man born blind, if he were instructed in mathematics, would be able to determine the visible figure of a body, when its real figure, distance, and position, are given. Dr. Saunderson understood the projection of the sphere, and perspective. Now, I require no more knowledge in a blind

man, in order to his being able to determine the visible figure of bodies, than that he can project the outline of a given body, upon the surface of a hollow sphere, whose centre is in the eye.

There is nothing surely to hinder a blind man from conceiving the position of the several parts of a body with regard to the eye, any more than from conceiving their situation with regard to one another; and therefore I conclude, that a blind man may attain a distinct conception of the visible figure of bodies.[12]

Since visible figure may be specified in terms of real figure, distance, position and projection, these are the only notions required for a distinct conception of this quality. It is not essentially linked to sensations or organs by which it is perceived or even to its being perceived at all. The blind person can understand its nature. No knowledge by acquaintance necessary.

At the same time, Reid agrees with Russell and the others that there is something that the perceptually impaired person does not understand regarding normal human perceptions of secondary and certain other qualities, like visible figure. Most obviously, the sighted person experiences sensations that the blind person does not. Plus, according to Reid, there are conceptual distinctions between the blind and sighted person. First, the blind person does not associate visible figure with colour, but a sighted perceiver hardly ever perceives visible figure without colour. Reid says that colour and visible figure for most perceivers are so closely linked that they have difficulty disjoining them in their imaginations. Moreover, the blind person learns about visible figure by reasoning from mathematical principles, whereas the sighted person 'has it presented to his eye at once, without any labour, without any kind of reasoning, by a kind of inspiration'. Finally, while the blind person considers a thing's visible figure only by making inferences based on its real figure and relative distance and position, the sighted person uses visible figure as a natural sign for real figure.[13] These differences, which concern the blind and sighted subjects' mental operations, however, do not bear on Reid's identification of visible figure with the projection of a thing's real figure. The identity holds.

While acknowledging that there may be a variety of conceptual and otherwise cognitive distinctions between perceptually impaired and perceptually able subjects, a follower of Reid can still say that secondary quality sensations do not feature in conceptions of secondary qualities' physical natures. Secondary qualities aren't essentially tied to sensations. Reid's theory offers a framework for identifying secondary qualities with the physical causes of sensations, not the sensations themselves, as Russell and others do.

Notes

1. Robinson, *Perception*, p. 63.
2. Key contributions to this debate include Block and Stalnaker, 'Conceptual Analysis, Dualism and the Explanatory Gap'; Chalmers and Jackson, 'Conceptual Analysis and Reductive Explanation'. Chalmers and Jackson favour the need for an a priori connection, but, if Block and Stalnaker are correct that such a connection is not necessary, then they offer one more reason for thinking that Robinson's objection fails.
3. Reid, *Essays on the Intellectual Powers of Man*, 2.3/80, 2.6/102.
4. Reid, *Essays on the Intellectual Powers of Man*, 2.3/84.
5. Reid, *Essays on the Intellectual Powers of Man*, 2.3/85.
6. Reid, *Essays on the Intellectual Powers of Man*, 2.17/209.
7. Reid, *Inquiry into the Human Mind*, 2.8/39.
8. Russell, *The Problems of Philosophy*, pp. 12–16. Russell dismisses colours as qualities of physical objects on the basis of their supposed superfluousness.
9. Fisher, 'Pierre Gassendi'.
10. Broad, 'The Theory of Sensa', pp. 126–7.
11. Foster, *The Nature of Perception*, pp. 131–2.
12. Reid, *Inquiry into the Human Mind*, 6.7/96.
13. Reid, *Inquiry into the Human Mind*, 6.7/97–8.

9

A Historical Objection

The final objection to the identity theory of secondary qualities, the resemblance objection, merits attention on historical
grounds. The resemblance objection depends on two commitments for which today's philosophers exhibit relatively little
sympathy. The first is the atomic or corpuscular theory of matter,
which in the seventeenth and eighteenth centuries was rapidly
replacing its Scholastic predecessors. The second is the view that
immediate objects of perception are mental. That is, the resemblance objection assumes the falsity of Direct Realism and the
truth of either Indirect Realism or Idealism. In the context of
this discussion, such a starting point is question begging. But the
resemblance objection represents the most prominent argument
for the mentality of secondary qualities from Reid's day and one
with which he is deeply concerned, because of its conceptual links
to the Way of Ideas. The following pages attempt to reconstruct
the resemblance objection in accordance with Reid's analysis of
the Way of Ideas and to highlight the features of his doctrine of
primary and secondary qualities that bear on his response to that
tradition.

According to Reid, adherents of the Way of Ideas include
Platonists, Aristotelians, Epicureans, Cartesians and Lockeans,
the only noted exception being Antony Arnauld.[1] Reid holds the
Way of Ideas responsible for the external world scepticism of
Berkeley and Hume.[2] Among Scholastics, the resemblance theory
manifests itself in terms of physical causation by contagion of
forms.[3] Just as they explain heat transfer by saying that one warm
body produces a form of warmness in another, so in perception a
red object transmits a 'sensible form' of redness to the eyes and
mind. The phenomenological experience is of the same form as

the physical quality. So generations of philosophers rely on the Way of Ideas to account for the intentionality and epistemic accuracy of perceptions or, more accurately, of sensations. The sensible form impressed on the mind resembles the object of perception in virtue of sharing its formal cause. And this resemblance makes the resulting idea or sensible form both literally informative and *of* the object.

According to Reid, early moderns take the immediate objects of all thought to be mental entities, ideas or images, following the Scholastic model of sensible forms.[4] In veridical perceptions or conceptions of external objects, the subject perceives an idea that resembles the external object. Descartes describes resemblance as analogous to the resemblance between an engraving and the thing of which it is an engraving.[5] A circular object, for example, may be represented an ellipse, but not a square. Although the circle and ellipse are not the same shape, one resembles the other well enough for perception by resemblance to ideas. For these philosophers, as for the Scholastics, the accuracy and intentionality of perceptions depend on the resemblance between object and idea or sensation.

Whether you find a resemblance between sensations and physical properties of depends on what sorts of properties you take physical objects to have. Many early moderns understand the physical world as being composed of tiny particles with only size, shape, position and motion.[6] These combine to form objects with certain complex qualities like hardness and solidity. But it seems that no combination of size, shape and position could ever form a quality that resembles naive ideas, Cartesian images or Reidian sensations.

Consider the resemblance objection as deriving from the Way of Ideas and early modern atomism as follows.[7] First, the Way of Ideas regards mental entities, ideas or sensations as immediate objects of thought. A physical property or object becomes an object of thought only by resembling some sensation or idea. However, scientific, physical properties do not include anything resembling secondary quality sensations. So secondary qualities, which are certainly objects of thought, can't be physical properties.

Reid agrees that sensations don't resemble physical properties. But, overall, Reid finds this case unconvincing. According to him, although philosopher after philosopher defends non-resemblance between ideas and physical properties, the contention that perceptions require resemblance between sensation and property borders on silly. In addition to yielding its unlikely conclusion, Reid thinks, the rule is vulnerable to counterexamples, in particular from perceptions of primary qualities.

Reid accuses his predecessors of asserting non-resemblance between secondary quality sensations and physical properties without defending the requirement for such a resemblance. The criticism seems apt. Galileo, Descartes and Locke all have passages in which they move immediately from the claim that secondary quality sensations or properties resembling them are not possessed by physical objects to the conclusion that perceivers mistakenly attribute secondary qualities to mind-external physical objects. Moreover, these writers treat the observation that physical properties don't resemble secondary quality sensations as if it were the locus of disagreement between common sense and their theories of secondary qualities.

For example, although it does not mention resemblance explicitly, Galileo's *The Assayer* contains a thought experiment to show that sensations may be distinct in kind from physical properties. Galileo asks his reader to consider what it is like to have a hand tickled with a feather. The feather moves about your skin, and you have certain phenomenological experiences, a tickling sensation. Galileo differentiates the sensation from the physical features of the feather and your hand, 'This titillation belongs entirely to us and not to the feather'. Then, he applies the distinction to secondary qualities explicitly, saying, 'I believe that no more solid an existence belongs to many qualities which we have come to attribute to physical bodies – tastes, odours, colours, and many more'. By thinking of secondary qualities as possessed by physical objects, he says, many have mistakenly attributed secondary quality sensations to physical objects: 'Many sensations which are supposed to be qualities residing in external objects have no real existence save in us, and outside ourselves are mere names'.[8] For Galileo, secondary qualities cannot be identical to physical properties, because physical objects cannot possess secondary quality sensations. However, the pages that contain this argument do not say why physical objects need to possess sensations in order to possess secondary qualities. So Galileo treats the resemblance requirement as given and the fact of non-resemblance as a discovery that overturns common sense understandings of secondary qualities.

Descartes says a bit more about non-resemblance. He claims that, just as words do not resemble the objects to which they refer, secondary quality ideas do not resemble their causes. Even primary qualities, according to Descartes, do not resemble their corresponding sensations in every respect.[9] This leads him to conclude, like Galileo, that secondary qualities are really just sensations posing as qualities.[10] Consider a secondary quality – like colour – as a sensation. There is no problem understanding it. It is 'clear' and 'distinct'. But as a quality of a physical body, this colour idea makes about as much sense as pain would. You can't meaningfully ascribe it to a non-sentient object. Other qualities, like shape and motion, clearly fit as properties of physical objects. Thus, Descartes, like Galileo, holds that secondary qualities are merely mental, because secondary quality sensations differ in kind from physical properties. And, like Galileo, Descartes seems to take for granted in these passages that resemblance or likeness in kind between secondary quality sensations and secondary qualities is required to perceive them.

Locke is even more explicit about resemblance between idea and quality. He extends Descartes's claims, proposing a primary/secondary quality division based on the resemblance of physical qualities to ideas. He writes,

> The ideas of primary Qualities of Bodies, are Resemblances of them, and their Patterns do really exist in the Bodies themselves; but the Ideas, produced in us by these Secondary Qualities, have no resemblance of them at all. There is nothing like our Ideas, existing in the Bodies themselves.[11]

Primary qualities, according to Locke, are somehow like the sensations that they cause. Secondary qualities are not. This is why Locke characterises secondary qualities as 'nothing but powers' and 'powers barely'.[12] He acknowledges that secondary quality sensations issue from external, physical properties. But he won't say that these causal properties just are the secondary qualities, since nothing about them resembles the objects of secondary quality ideas. Locke offers some examples involving heat:

> [Secondary] Qualities are commonly thought to be the same in those Bodies, that those Ideas are in us, the one the perfect resemblance of the other, as they are in a Mirror; and it would by most Men be judged very extravagant, if one should say otherwise. And yet he, that will consider, that the same Fire, that at one distance produces in us the Sensation of Warmth, does at a nearer approach, produce in us the far different Sensation of Pain, ought to bethink himself, what Reason he has to say, That his Idea of Warmth, which was produced

in him by the Fire, is actually in the Fire; and his Idea of Pain, which the Fire produced in him the same way, is not in the Fire.[13]

Just like Galileo and Descartes, Locke believes that attributing warmth to fire mistakenly ascribes a secondary quality sensation to a physical object. Another passage:

> Though receiving the Idea of Heat, or Light, from the Sun, we are apt to think, 'tis a Perception and Resemblance of such a Quality in the Sun: yet when we see Wax, or a fair Face, receive change of Colour from the Sun, we cannot imagine, that to be the Reception or Resemblance of any thing in the Sun, because we find not those different Colours in the Sun it self.[14]

For Locke, colour, smell and sound ideas exist primarily in the minds of perceivers, as immediate objects of perceptions. But these ideas do not resemble anything mind-external. Thus, he concludes that secondary qualities are only ideas in minds, even if physical objects possess 'bare powers' that cause the secondary quality sensations. Once again, the philosopher moves directly from the fact that secondary quality sensations do not resemble physical properties to the conclusion that secondary qualities are not possessed by, and therefore are not identical to, physical properties.

In each thinker, the contention that secondary qualities are possessed by physical objects magically morphs into a different claim, that physical objects possess something resembling secondary quality sensations or, even more radical, that physical objects possess human sensations. Perhaps Galileo holds the latter view. If so, his sentiment reappears with David Hume, who says that human perceivers follow a 'blind and powerful instinct of nature' believing that sensory images are external objects. On the other hand, Hume says that such an opinion gets 'destroyed by the slightest philosophy', because in fact human minds can only engage mental images.[15] So Hume also endures the rule that there are no physical properties that resemble sensations of secondary qualities. And he assumes that thoughts of physical properties or objects involve resemblances with sensations.

Reid concurs with all of these writers about the non-resemblance between sensations and physical properties. But, contrary to them, he thinks that every sane person also concurs on this point. Physical properties are not like sensations. Why treat it as a grand discovery? If you did not distinguish properties and sensations, wouldn't you confuse perceivable properties with the sensations they cause? But in fact you're probably pretty good at separating mental and physical, seeming and being. You might even say things that demonstrate a ready awareness of the difference, like, 'The baby feels warm, but according to the thermometer she isn't running a fever.' Galileo's tickling example capitalises on just this point, that perceivers do not confuse sensations with physical properties. So Reid defends non-resemblance as the common sense view, even with regard to secondary qualities:

> The vulgar are commonly charged by philosophers, with the absurdity of imagining the smell in the rose to be something like to the sensation of smelling: but I think, unjustly; for they neither give the same epithets to both, nor do they reason in the same manner from them. We say, This body smells sweet, that stinks; but we do not say, This mind smells sweet and that stinks. Therefore smell in the rose, and the sensation which it causes, are not conceived, even by the vulgar, to be things of the same kind, altho' they have the same name.[16]

Reid is seconded by twentieth-century writers P. F. Strawson and Wilfred Sellars.[17] Moreover, the English language favours secondary qualities as real, physical and not identical to sensations. For example, sensations are said to disappear when no longer felt, but colours, smells and sounds do not. The Way of Ideas, by treating non-resemblance as unexpected, predicts confusion where there is none. Reid sees nothing scandalous in the lack of resemblance between sensations and physical properties.

He finds the requirement for resemblance to be false. The requirement supposes that all immediate objects of thought are ideas or sensations. Physical objects and properties become mediate objects of thought only by resembling immediately perceived or conceived ideas. To think about a physical object, then, two sub-relations must hold: the thought's being of a sensation and the sensation's resembling some object.[18] Without resemblance between sensation and physical object, thoughts never get beyond sensations. Whenever sensations happen to resemble physical properties or objects, thoughts about the sensations are also indirectly about the properties and objects.

Reid responds that he regularly conceives of primary qualities, which are physical properties, even though they aren't sensations nor do they resemble sensations. Conceptions of primary quality sensations are not conceptions of primary qualities, and conceptions of primary qualities are not conceptions of primary quality sensations. Reid's argument against the resemblance requirement forms the crux or 'experimentum crucis' of his *Inquiry*. He issues his challenge,

> Extension, figure, motion, may, any one, or all of them, be taken for the subject of this experiment. Either they are ideas of sensation, or they are not. If any one of them can be shown to be an idea of sensation, or to have the least resemblance to any sensation, I lay my hand upon my mouth, and give up all pretence to reconcile reason to common sense in this matter, and must suffer the ideal scepticism to triumph. But if, on the other hand, they are not ideas of sensation, nor like to any sensation, then the ideal system is a rope of sand, and all the labored arguments of the sceptical philosophy against a material world, and against the existence of every thing but impressions and ideas, proceed upon a false hypothesis.[19]

All Reid needs to reject the resemblance rule is a single case where an idea or conception of a physical property is not an idea of a sensation. He names three: extension, figure and motion. If his analysis of primary quality perceptions is correct, he could add three more: hardness, solidity and divisibility. These, Reid says, should settle the matter.

Reid agrees with Galileo, Descartes and Locke that physical properties do not resemble secondary quality sensations. But Reid goes even further. Following Berkeley, Reid holds that no physical property resembles any sensation, nor could it.[20] If perceived properties have to resemble sensations, then, Reid thinks, it is impossible to perceive or even think about external objects. Nothing in the physical world satisfies the resemblance requirement.

Moreover, if the resemblance rule were to hold, then all supposed conceptions of physical properties are really conceptions of sensations. Surely this is not the case. Reid observes that sensations accompany perceptions of primary qualities, just as they do secondary qualities.[21] As mentioned in Part II, the difference in primary and secondary quality sensations arises from their roles as natural signs. Primary quality sensations are type-3 natural signs that trigger direct and distinct conceptions, whereas secondary quality sensations are type-1 signs that initially give only relative and obscure conceptions. Reid admits that it is difficult to attend to primary quality sensations, but, having overcome this difficulty, he asks his reader to compare conceptions of primary qualities to conceptions of sensations caused by them.[22] Are these conceptions in fact the same? Do they take the same objects? Reid asserts that they are manifestly distinct.

Reid's clearest treatment of this experiment deals with a conception of hardness and a conception of the sensation caused by touching a hard object.[23] Again, the Way of Ideas says that these are the same conception. But is this in fact the case? It seems that the answer is no. To make the point more vividly, Reid considers an extreme case: 'If a man runs his head with violence against a pillar, I appeal to him, whether the pain he feels resembles the hardness of the stone; or if he can conceive any thing like what he feels to be in an inanimate piece of matter'. The one who runs his head into a pillar certainly has a sensation, pain, and a little introspection gives him a conception that has the pain as its referential content. But the referential content of the man's conception of the pillar's hardness is of the firm cohesion of the pillar's parts if it is a conception of anything, not of a pain.[24] The man forms two distinct conceptions, one of the pain sensation, a mental operation, and another of the pillar's hardness, a physical property. His thought of the pillar is not mediated by thoughts of ideas. The Way of Ideas and the resemblance rule are false.

For good measure, Reid makes similar assertions concerning conceptions of other primary qualities, finally remarking,

> But let us, as becomes philosophers, lay aside authority; we need not surely consult Aristotle or Locke, to know whether pain be like the point of a sword. I have as clear a conception of extension, hardness, and motion, as I have of the point of a sword; and, with some pains and practice, I can form as clear a notion of the other sensations of touch, as I have of pain.[25]

Human perceivers seem perfectly able to accurately conceive of physical properties, even when they bear no resemblance to sensations.

The Way of Ideas cannot cope with conceptions that are not, first and foremost, of ideas or sensations. It says that conceptions of primary qualities just are conceptions of primary quality sensations. But this is not true. Perceivers have conceptions for each of the primary qualities and, if you attend to them, conceptions for the sensations that they cause. As Reid announces in triumph, 'The very conception of [primary qualities] is irreconcilable to the principles of all our philosophic systems of understanding'.[26] And again, 'The very existence of our conceptions of extension, figure, and motion, since they are neither ideas of sensation nor reflection, overturns the whole ideal system'.[27]

Primary qualities do not resemble the sensations that trigger conceptions of them. So Reid finds a reason to dismiss the general rule that external properties must resemble sensations. The Way of Ideas offers no compelling reason to say that secondary qualities, if they are physical, must resemble sensations. Perhaps, they cause sensations, and one may conceive of them as sensation-causing but otherwise unknown properties of physical objects. If so, the identity theory of secondary qualities remains unharmed by this objection.

Notes

1. Reid, *Essays on the Intellectual Powers of Man*, 2.13/165–70.
2. Reid, *Essays on the Intellectual Powers of Man*, 1.1/31, 2.8/112, 2.9/132. For a more detailed account of Reid on the restrictions imposed by the Way of Ideas, see Greco, 'Reid's Critique of Berkeley and Hume'.
3. Reid, *Essays on the Intellectual Powers of Man*, 2.8/112–15.
4. Reid, *Essays on the Intellectual Powers of Man*, 2.8/115–16.
5. Descartes, *Optics*, p. 165.
6. Adams, 'Editor's Introduction', pp. xiv.
7. This section owes much to Greco, 'Reid's Critique of Berkeley and Hume'.
8. Galileo, *The Assayer*, pp. 275–7.
9. Descartes, *Optics*, pp. 165–7.
10. Descartes, *Principles of Philosophy*, 1.71/219.
11. Locke, *An Essay Concerning Human Understanding*, 2.8.15/137.
12. Locke, *An Essay Concerning Human Understanding*, 2.8.24/141.
13. Locke, *An Essay Concerning Human Understanding*, 2.8.16/137.
14. Locke, *An Essay Concerning Human Understanding*, 2.9.25/142.
15. Hume, *Enquiry Concerning the Human Understanding*, 12.1/104.
16. Reid, *Inquiry into the Human Mind*, 2.9/42.
17. Strawson, 'Perception and its Objects', pp. 48–50; Sellars, 'Foundations for a Metaphysics of Pure Process'; Sellars, 'Empiricism and the Philosophy of Mind'.
18. Reid, *Inquiry into the Human Mind*, 5.8/73.
19. Reid, *Inquiry into the Human Mind*, 5.7/70.
20. Berkeley, *Three Dialogues*, p. 79; Reid, *Inquiry into the Human Mind*, 5.8/75.
21. Reid, *Inquiry into the Human Mind*, 5.5/63–4.

22. Reid, *Inquiry into the Human Mind*, 5.2/56–7.
23. Reid, *Inquiry into the Human Mind*, 5.2/55.
24. Reid, *Inquiry into the Human Mind*, 5.4/61.
25. Reid, *Inquiry into the Human Mind*, 5.7/69.
26. Reid, *Inquiry into the Human Mind*, 5.7/67.
27. Reid, *Inquiry into the Human Mind*, 5.7/69.

10

Conclusion

Direct Realism is worth defending. Its common sense appeal and epistemic optimism affirm humanity's moral responsibility, preserve its confidence in scientific investigation, and satisfy its intuitions about phenomenology and the objective physical world. Although the Problem of Secondary Qualities presents a formidable obstacle to accepting Direct Realism, there are Direct Realist accounts of perception and theories of secondary qualities, like Thomas Reid's, that can overcome it.

The Problem of Secondary Qualities infers the falsity of Direct Realism from three premises: a spreading principle, that a good analysis of secondary qualities should extend to perception generally (SP); an empirical claim, that secondary qualities are among the objects of human perception (OC); and a supposed scientific discovery, that physical objects do not possess secondary qualities (NPT). From these, it concludes that no perceivable qualities are possessed by physical objects, which would make Direct Realism false.

All three premises from the Problem of Secondary Qualities are prima facie plausible. But while SP and OC hold up against a wide range of criticisms, NPT is somewhat weaker. It rests on a poorly defended hypothesis about whether secondary qualities are physically causal. They are not, it is supposed, because scientific properties account for the whole of physical causation, particularly between the objects of perceptions and human sense organs, making additional properties superfluous. But scientists and philosophers who say that physical objects do not possess secondary qualities because they are non-causal have spoken too soon. Even if causal-scientific properties formed the entire causal story about stimulating sense organs, this does not bar secondary qualities from the world of science. They can be objective and causal if they are identical to known scientific properties.

One way to make these a posteriori identifications is on the basis of Thomas Reid's theory of secondary qualities. According to Reid, secondary qualities are physical properties and objects of human perceptions. But original conceptions of them are relative and obscure. You do not understand the natures of secondary qualities merely by perceiving them. Rather, you grasp them in terms of their relations to your sensations, which are natural signs for the qualities. The epistemic difficulty with these signs, as opposed to certain other natural signs, is that it is up to individuals to decipher the connections between the signs and the things that they signify. If a certain sensation signifies the smell of a rose, then you must learn to associate the representative smell with the represented rose through experience and scientific investigation. Further investigative exercises eventually may offer you an understanding of each secondary quality as it is in itself.

Reid's identity theory of secondary qualities proves defensible against four formidable objections: non-correspondence, a priori sufficiency, knowledge by acquaintance, and resemblance. Each of these either wrongly conflates sensations with perceptions or fails to acknowledge the obscure and relative nature of unlearned secondary quality perceptual conceptions.

This is not to say that there can be no correlations between secondary qualities and the sensations that they cause. Heat sensations certainly depend on the temperature of the tactilely perceived object, even if they also depend on the object's heat capacity and the physiological features of the perceiver's skin. Cain Todd's *Philosophy of Wine* offers an argument for the objectivity of certain gustatory properties on the basis of correlations between taste sensations and chemical features of the wine. Perhaps 'elegant', 'overripe' and 'charming' pick out natural kinds of properties intrinsic to wine, despite the fact that wine tasters conceive of these properties in sensation-relative ways. And Byrne and Hilbert propose a method by which a scientifically sophisticated perceiver might be able to relate colour appearances to the spectral reflectances of visible objects.[1] If you have the right background beliefs about your own sensations, your own physiology and the physics of light, you might be able to infer something about the relative stimulations of your retinal cones from your sensations and, from there, something more about the kinds of light causing those stimulations. So perhaps you have the capacity to discover something about the nature of a thing's colour by looking at it.

Reid's theory of secondary qualities draws attention to these identity theories in order to discover the connections between sensations and their causes. But, on Reid's version, the outcomes of such investigations have no bearing on whether any particular secondary quality is physical or objective. Secondary quality sensations arise because scientific properties stimulate sense organs, so secondary qualities can only be scientific, no matter the type–type correspondences between them and sensations. Sensations are not secondary qualities, and since original conceptions of secondary qualities do not say that there must be an invertible function from sensation type to quality type, there is no reason to expect to find such a function, outside of wishful thinking. The investigation is open.

The cost to Reid's approach turns out to be traditional Western colour categories, since they don't reflect natural kinds. Contrary to what many have supposed, 'red', 'blue' and 'green' don't name objective colour types, because they do not pick out single unified properties of physical bodies. This is not a problem, however, since our insistence on traditional colour categories turns out to be unnecessarily ethnocentric. Giving it up is a small price for maintaining common sense and Direct Realism.

Other objections bar the identification between secondary qualities and physical properties for philosophical reasons. One says that physical properties cannot explain the phenomenological effects of secondary qualities. Another says that physical properties can be known by the perceptually impaired but secondary qualities cannot, so the first cannot be identified with the second. These objections, however, confuse sensations with their causes. To understand the natures of secondary qualities, it's not the contingent causal connections to sensations that drive the investigation. While the perceptually impaired may lack a certain sensation-relative understanding of secondary qualities, this in no way inhibits them from understanding their natures.

Finally, the early modern objection to the identity theory of secondary qualities, based on the Way of Ideas, demands that conceptions of physical properties be mediated by sensations that resemble the properties. But the Way of Ideas asks too much. If the resemblance theory were correct, then conceptions of primary qualities would be of properties that resemble certain sensations.

And they clearly aren't. The point of a sword is nothing like the pain that it causes, nor is a conception of the sword identical to a conception of the pain.

As a group, these objections to the identity theory have an internal inconsistency as well. The non-correspondence objection conflicts with the other three. The a priori sufficiency, knowledge by acquaintance, and resemblance arguments are all non-scientific or even a priori objections to the identity theory. They say that such a theory is not even conceivable. But the non-correspondence objection is based on supposed scientific findings concerning the relations of your sense experiences and perceiver-independent properties of physical objects. So why would scientists investigate an inconceivable hypothesis? If science has discovered that secondary qualities are not identifiable with scientific properties, then the theory is conceivable. On the other hand, if the theory is inconceivable, then the scientific findings are pointless. Thus, as P. M. S. Hacker points out, if someone wished to object to the identity theory, she could attempt it on the basis of the non-correspondence objection or an a priori objection, but not both.[2] So these objections were dubious even before considering them individually.

On balance, Reid's theory of secondary qualities offers a highly plausible, dialectically defensible response to the Problem of Secondary Qualities.

Notes

1. Byrne and Hilbert, 'Color Realism and Color Science', p. 14.
2. Hacker, *Appearance and Reality*, p. 141.

Bibliography

Adams, Robert, 'Editor's Introduction', in George Berkeley, *Three Dialogues between Hylas and Philonous* (Indianapolis: Hackett, 1979), pp. xi–xxvi.

Allen, Keith, 'Inter-Species Variation in Colour Perception', *Philosophical Studies* 142, no. 2 (2009): 197–220.

Allen, Keith, 'Revelation and the Nature of Colour', *Dialectica* 65, no. 2 (2011): 153–76.

Armstrong, David M., *Belief, Truth and Knowledge* (Cambridge: Cambridge University Press, 1973).

Armstrong, David M., *A Materialist Theory of the Mind* (New York: Psychology Press, 1993).

Ayer, A. J., *Central Questions of Philosophy* (London: Weidenfeld & Nicolson, 1973).

Ayer, A. J., *The Foundations of Empirical Knowledge* (London: Macmillan, 1963).

Ayer, A. J., 'Has Austin Refuted the Sense-Datum Theory?' *Synthese* 18 (1967): 117–40.

Bear, Mark F., Barry W. Connors and Michael A. Paridiso, *Neuroscience: Exploring the Brain*, 3rd edn (New York: Lippincott Williams & Wilkins, 2007).

Bennett, Jonathan, *Locke, Berkeley, Hume* (Oxford: Clarendon Press, 1971).

Bennett, Jonathan, 'Substance, Reality, and Primary Qualities', *American Philosophical Quarterly* 2 (1965): 1–17.

Berkeley, George, *Three Dialogues between Hylas and Philonous* (Indianapolis: Hackett, [1713] 1979).

Berkeley, George, *A Treatise Concerning the Principles of Human Knowledge* (New York: Oxford University Press, [1710] 1998).

Berns, Roy S., Fred W. Billmeyer and Max Saltzman, *Billmeyer and Saltzman's Principles of Color Technology*, 3rd edn (New York: John Wiley & Sons, 2000).

Block, Ned and Robert Stalnaker, 'Conceptual Analysis, Dualism and the Explanatory Gap', *Philosophical Review* 108 (1999): 1–46.

Boswell, James, *Boswell's Life of Johnson* (New York: Scribner, [1791] 1945).

Broad, C. D., 'The Theory of Sensa', in Robert J. Swartz (ed.), *Perceiving, Sensing, and Knowing* (Los Angeles: University of California Press, 1965), pp. 85–129.

Buras, J. Todd, 'The Function of Sensations in Reid', *Journal of the History of Philosophy* 47 (2009): 329–55.

Buras, J. Todd, 'The Nature of Sensations in Reid', *History of Philosophy Quarterly* 22 (2005): 221–38.

Buras, J. Todd, 'The Problem with Thomas Reid's direct realism', in John Haldane and Stephen Read (eds), *The Philosophy of Thomas Reid: A Collection of Essays*, (Malden, MA: Blackwell Pub, 2003), pp. 44–64.

Byrne, Alex and David R. Hilbert, 'Color Realism and Color Science', *Behavioral and Brain Sciences* 26 (2003): 1–44.

Callergard, Robert, *An Essay on Thomas Reid's Philosophy of Science* (Stockholm: Stockholm University, 2006).

Chalmers, Alan, 'Atomism from the 17th to the 20th Century', in Edward N. Zalta (ed.), *The Stanford Encyclopedia of Philosophy* (Winter 2014 Edition), <http://plato.stanford.edu/entries/atomism-modern/>, accessed 4 October 2016.

Chalmers, David and Frank Jackson, 'Conceptual Analysis and Reductive Explanation', *Philosophical Review* 110 (2001): 315–61.

Chirimuuta, Mazviita, *Outside Color: Perceptual Science and the Puzzle of Color in Philosophy* (Cambridge, MA: The MIT Press, 2015).

Churchland, Paul, 'Eliminative Materialism and the Propositional Attitudes', *Journal of Philosophy* 78 (1981): 67–90.

Clark, Austen, *A Theory of Sentience* (New York: Oxford University Press, 2000).

Cohen, Jonathan, *The Red and the Real* (New York: Oxford University Press, 2009).

Copenhaver, Rebecca, 'Thomas Reid's Direct Realism', *Reid Studies* 4, no. 1 (2000): 17–34.

Cosmides, Leda and John Tooby, 'Forward', in Simon Baron-Cohen (ed.), *Mindblindness* (Cambridge, MA: MIT Press, 1995), pp. xi-xviii.

Dancy, Jonathan, 'Introduction', in George Berkeley, *A Treatise Concerning the Principles of Human Knowledge* (New York: Oxford University Press, 1998), pp. 5–69.

Dennett, Daniel, *Consciousness Explained* (New York: Bay Back Books, 1991).

Descartes, Rene, *Meditations*, in John Cottingham, Robert Stoothoff and Dugald Mudroch (trans.), *The Philosophical Writings of Descartes*, vol. 2 (New York: Cambridge University Press, [1641] 1985), pp. 11–62.

Descartes, Rene, *Optics*, in John Cottingham, Robert Stoothoff and Dugald Mudroch (trans.), *The Philosophical Writings of Descartes*, vol. 1 (New York: Cambridge University Press, [1637] 1985), pp. 152–75.

Descartes, Rene, *Principles of Philosophy*, in John Cottingham, Robert Stoothoff and Dugald Mudroch (trans.), *The Philosophical Writings of Descartes*, vol. 1 (New York: Cambridge University Press, [1644] 1985), pp. 177–291.

Dretske, Fred, *Naturalizing the Mind* (Cambridge, MA: MIT Press, 1995).

Dretske, Fred, *Perception, Knowledge and Belief* (New York: Cambridge University Press, 2000).

Dretske, Fred, 'Perception without Awareness', in Tamar Gendler and John Hawthorne (eds), *Perceptual Experience* (New York: Oxford University Press, 2006), pp. 147–80.

Elliot, Andrew and Daniela Niesta, 'Romantic Red: Red Enhances Men's Attraction to Women', *Journal of Personality and Social Psychology* 95 (2008): 1150–64.

Falkenstein, Lorne and Giovanni Grandi, 'The Role of Material Impressions in Reid's Theory of Vision: A Critique of Gideon Yaffe's "Reid on the Perception of the Visible Figure"', *The Journal of Scottish Philosophy* 1 (2003): 117–33.

Fish, William, *Philosophy of Perception: A Contemporary Introduction* (New York: Routledge, 2010).

Fisher, Saul, 'Pierre Gassendi', in Edward N. Zalta (ed.), *The Stanford Encyclopedia of Philosophy* (Spring 2014 Edition), <http://plato.stanford.edu/archives/win2009/entries/gassendi/>, accessed 6 October 2016.

Foster, John, *The Nature of Perception* (New York: Oxford University Press, 2000).

Galileo, *The Assayer*, in Stillman Drake (trans.), *Discoveries and Opinions of Galileo* (New York: Anchor Books, [1623] 1957), pp. 229–80.

Greco, John, 'Reid's Critique of Berkeley and Hume: What's the Big Idea?' *Philosophy and Phenomenological Research* 40 (1995): 279–96.

Greco, John, 'Reid's Reply to the Skeptic', in Terence Cuneo and René van Woudenberg (eds.), *The Cambridge Companion to Thomas Reid* (New York: Cambridge University Press, 2004), pp. 134–55.

Hacker, Peter, *Appearance and Reality* (Oxford: Basil Blackwell, 1989).

Hanson, N. R., *Patterns of Discovery* (Cambridge University Press, 1958).

Hardin, C. L., *Color for Philosophers: Unweaving the Rainbow* (Indianapolis: Hackett, 1988).

Hardin, C. L., 'A Spectral Reflectance Doth Not a Color Make', *Journal of Philosophy* 100 (2003): 191–202.

Hart, N. S., J. C. Partridge, A. T. D. Bennett and I. C. Cuthill, 'Visual Pigments, Cone Oil Droplets and Ocular Media in Four Species of Estrildid Finch', *Journal of Comparative Physiology A* 186 (2000): 68–94.

Hilbert, David R., 'What is Color Vision?', *Philosophical Studies* 68 (1992): 351–70.

Huemer, Michael, 'Sense-Data', in Edward N. Zalta (ed.), *The Stanford Encyclopedia of Philosophy* (Spring 2011 Edition), <http://plato.stanford.edu/entries/sense-data/> accessed 6 October 2016.

Huemer, Michael, *Skepticism and the Veil of Perception* (New York: Rowman & Littlefield, 2001).

Hume, David, *An Enquiry Concerning Human Understanding* (Indianapolis: Hackett, [1748] 1977).

Hume, David, *Treatise on Human Nature* (New York: Oxford University Press, [1739] 2000).

Ivery, Alfred L., 'The Ontological Entailments of Averroes' Understanding of Perception', in Simo Knuuttila and Pekka Karkkainen (eds), *Theories of Perception in Medieval and Early Modern Philosophy* (n.p.: Springer, 2008), pp. 73–86.

Jackson, Frank, *Perception: A Representative Theory* (New York: Cambridge, 1977).

Karkkainen, Pekka, 'Objects of Sense Perception in Late Medieval Erfurtian Nominalism', in Simo Knuuttila and Pekka Karkkainen (eds), *Theories of Perception in Medieval and Early Modern Philosophy* (n.p.: Springer, 2008), pp. 187–202.

Kripke, Saul, *Naming and Necessity* (Cambridge, MA: Harvard University Press, 1980).

Landesman, Charles, *Color and Consciousness: An Essay in Metaphysics* (Philadelphia: Temple University Press, 1989).

Laurence, Stephen and Eric Margolis, 'Abstraction and the Origin of General Ideas', *Philosophers' Imprint* 12, no. 19 (2012): 1–22.

Lehrer, Keith, *Knowledge* (New York: Clarendon Press, 1978).

Lehrer, Keith, 'Reid on Primary and Secondary Qualities', *The Monist* 61 (1978): 184–91.

Lehrer, Keith and Joseph Tolliver, 'Tropes and Truth', in Anne Reboul (ed.), *Philosophical Papers Dedicated to Kevin Mulligan* (2011), <http://www.philosophie.ch/kevin/festschrift/>, accessed 5 October 2016.

Lemos, Noah, *Common Sense: A Contemporary Defense* (New York: Cambridge University Press, 2004).

Lewis, David, 'Finkish Dispositions', *The Philosophical Quarterly* 47 (1997): 143–58.

Locke, John, *An Essay Concerning Human Understanding*, ed. Peter H. Nidditch (New York: Oxford University Press, [1690] 1975).

McGinn, Colin, *The Subjective View* (Oxford: Clarendon Press, 1983).

Mackie, J. L., 'Dispositions, Grounds and Causes', *Synthese* 34 (1977): 361–70.

McKitrick, Jennifer, 'A Case for Extrinsic Dispositions', *Australasian Journal of Philosophy* 81 (2003): 155–74.

McKitrick, Jennifer, 'Reid's Foundation for the Primary/Secondary Quality Distinction', in John Haldane and Stephen Read (eds), *The Philosophy of Thomas Reid: A Collection of Essays* (Malden, MA: Blackwell, 2003), pp. 65–81.

Martin, M. G. F., 'Beyond Dispute: Sense-Data, Intentionality and the Mind–Body Problem', in Tim Crane and Sarah Patterson (eds), *History of the Mind–Body Problem* (London: Routledge, 2000), pp. 195–231.

Menzies, Peter, 'Critical Notices of *Nature's Metaphysics* by Alexander Bird', *Analysis* 69 (2009): 769–78.

Moore, G. E., 'Proof of an External World', in Ernest Sosa and Jaegwon Kim (eds), *Epistemology: An Anthology*, 2nd edn (Malden, MA: Blackwell, 2008).

Nichols, Ryan, *Thomas Reid's Theory of Perception* (New York: Oxford University Press, 2007).

Nichols, Ryan and Gideon Yaffe, 'Thomas Reid', in Edward N. Zalta (ed.), *The Stanford Encyclopedia of Philosophy* (Winter 2016 Edition), <http://plato.stanford.edu/entries/reid/>, accessed 6 October 2016.

Ornstein, Robert and Richard Thompson, *The Amazing Brain* (Boston: Houghton Mifflin, 1984).

Paddle, Robert, *The Last Tasmanian Tiger: The History and Extinction of the Thylacine* (New York: Cambridge University Press, 2002).

Pelser, Adam, 'Belief in Reid's Theory of Perception', *History of Philosophy Quarterly* 27 (2010): 359–78.

Pitcher, George, *A Theory of Perception* (Princeton: Princeton University Press, 1971).

Pitson, Tony, 'Reid on Primary and Secondary Qualities', *Reid Studies* 5 (2001): 17–34.

Price, H. H., *Perception*, 2nd edn (London: Methuen, 1950).

Prior, Elizabeth W., Robert Pargetter and Frank Jackson, 'Three Theses about Dispositions', *American Philosophical Quarterly* 19 (1982): 251–7.

Putnam, Hilary, 'Meaning and Reference', *Journal of Philosophy* 70 (1973): 699–711.

Quine, Willard V., 'On What There Is', *The Review of Metaphysics* 2 (1948): 21–38.

Raven, Peter H., Ray F. Evert, and Susan E. Eichhorn, *Biology of Plants*, 6th edn (New York: W. H. Freeman and Company, 1999).

Reid, Thomas, *The Correspondence of Thomas Reid*, ed. Paul Wood (University Park: Pennsylvania State University Press, 2002).

Reid, Thomas, *Essays on the Active Powers of Man*, ed. Knud Haakonssen and James A. Harris (University Park: Pennsylvania State University Press, [1788] 2010).

Reid, Thomas, *Essays on the Intellectual Powers of Man*, ed. Derek R. Brookes (University Park: Pennsylvania State University Press, [1785] 2002).

Reid, Thomas, *An Inquiry into the Human Mind on the Principles of Common Sense*, ed. Derek R. Brookes (University Park: Pennsylvania State University Press, [1764] 1997).

Roberson, Debi, Jules Davidoff, Ian Davies and Laura Shapiro, 'Color Categories: Evidence for the Cultural Relativity Hypothesis', *Cognitive Psychology* 50 (2005): 378–411.

Robinson, Howard, *Perception* (New York: Routledge, 2001).

Rousseau, Jean-Jacques, *Emile: or On Education*, trans. Allan Bloom (New York: Basic Books, [1762] 1979).

Russell, Bertrand, *The Problems of Philosophy* (n.p.: Valde Books, [1912] 2009).

Sellars, Wilfred, 'Empiricism and the Philosophy of Mind', in *Science, Perception and Reality* (Atascadero: Ridgeview Publishing Co., [1956] 1991), pp. 127–96.

Sellars, Wilfred, 'Foundations for a Metaphysics of Pure Process', *The Monist* (1981): 3–90.

Shrock, Christopher A., 'Yellow is Not a Color', *Southwest Philosophical Studies* 34 (2012): 58–64.

Silberberg, Martin S., *Chemistry: The Molecular Nature of Matter and Change*, 2nd edn (St Louis: McGraw-Hill, 2000).

Smart, J. J. C., 'On Some Criticisms of a Physicalist Theory of Colors', in Alex Bryne and David R. Hilbert (eds), *Readings on Color, Volume 1: The Philosophy of Color* (Cambridge, MA: MIT Press, [1975] 1997), pp. 1–10.

Smith, A. D., *The Problem of Perception* (Cambridge, MA: Harvard University Press, 2002).

Soteriou, Matthew, 'The Disjunctive Theory of Perception', in Edward N. Zalta (ed.), *The Stanford Encyclopedia of Philosophy* (Winter 2016 Edition), <http://plato.stanford.edu/entries/perception-disjunctive/>, accessed 5 October 2016.

Strawson, P. F., 'Perception and its Objects', in G. F. Macdonald (ed.), *Perception and Identity: Essays Presented to A. J. Ayer, with His Replies* (Ithaca, NY: Cornell University Press, 1979), pp. 41–60.

Tebaldi, David A., 'Thomas Reid's Refutation of the Argument from Illusion', in Stephen F. Barker and Tom L. Beauchamp (eds), *Thomas*

Reid: Critical Interpretations (University City Science Center, 1976), pp. 25–34.

Todd, Cain, *The Philosophy of Wine: A Case of Truth, Beauty and Intoxication* (Ithaca, NY: McGill-Queen's University Press, 2010).

Tye, Michael, *Consciousness, Color, and Content* (Cambridge, MA: MIT Press, 2000).

Van Cleve, James, 'Reid on the Real Foundation of the Primary–Secondary Quality Distinction', in Lawrence Nolan (ed.), *Primary and Secondary Qualities: The Historical and Ongoing Debate* (New York: Oxford University Press, 2011), pp. 274–303.

Van Cleve, James, 'Reid's Theory of Perception', in Terence Cuneo and René van Woudenberg (eds), *The Cambridge Companion to Thomas Reid* (New York: Cambridge University Press, 2004), pp. 101–33.

Warnock, G. J., 'Seeing', in Robert J. Swartz (ed.), *Perceiving, Sensing, and Knowing* (Los Angeles: University of California Press, 1965), pp. 49–67.

Wolterstorff, Nicholas, *Thomas Reid and the Story of Epistemology* (New York: Cambridge University Press, 2001).

Wooding, Stephen, 'Phenylthiocarbamide: A 75-Year Adventure in Genetics and Natural Selection', *Genetics* 172 (2006): 2015–23.

Yaffe, Gideon, 'The Office of an Introspectible Sensation: A Reply to Falkenstein and Grandi', *The Journal of Scottish Philosophy* 1 (2003): 135–40.

Yaffe, Gideon, 'Reid on the Perception of Visible Figure', *The Journal of Scottish Philosophy* 1 (2003): 103–15.

Index